Geophysical and Hydrologic Analysis of an Earthen Dam Site in Southern Westchester County, New York

By Anthony Chu, Frederick Stumm, Peter K. Joesten, and Michael L. Noll

Prepared in cooperation with the
New York City Department of Environmental Protection

Scientific Investigations Report 2012–5247

U.S. Department of the Interior
U.S. Geological Survey

U.S. Department of the Interior
KEN SALAZAR, Secretary

U.S. Geological Survey
Suzette M. Kimball, Acting Director

U.S. Geological Survey, Reston, Virginia: 2013

For more information on the USGS—the Federal source for science about the Earth, its natural and living resources, natural hazards, and the environment, visit http://www.usgs.gov or call 1–888–ASK–USGS.

For an overview of USGS information products, including maps, imagery, and publications, visit http://www.usgs.gov/pubprod

To order this and other USGS information products, visit http://store.usgs.gov

Suggested citation:
Chu, Anthony, Stumm, Frederick, Joesten, P.K., and Noll, M.L., 2013, Geophysical and hydrologic analysis of an earthen dam site in southern Westchester County, New York: U.S. Geological Survey Scientific Investigations Report 2012–5247, 64 p. (Also available at http://pubs.usgs.gov/sir/2012/5247/.)

Acknowledgements

The authors thank Matthew Osit, Richard Zunino, George Schmitt, and Masud Ahmed of the New York City Department of Environmental Protection for their technical assistance and logistical support. The authors also thank William Capurso, Michael Como, Irene Fisher, Robert Pearsall, and Agnes Cwalina of the U.S. Geological Survey for their field support.

Contents

Figures

Tables

Conversion Factors and Datum

Multiply	By	To obtain
Length		
inch (in.)	2.54	centimeter (cm)
foot (ft)	0.3048	meter (m)
Area		
acre	0.004047	square kilometer (km^2)
Volume		
million gallons (Mgal)	3,785	cubic meter (m^3)
Flow rate		
gallon per minute (gal/min)	0.06309	liter per second (L/s)
Hydraulic conductivity		
foot per day (ft/d)	0.3048	meter per day (m/d)
Transmissivity*		
foot squared per day (ft^2/d)	0.09290	meter squared per day (m^2/d)

Temperature in degrees Celsius (°C) may be converted to degrees Fahrenheit (°F) as follows:

$$°F = (1.8 \times °C) + 32$$

Temperature in degrees Fahrenheit (°F) may be converted to degrees Celsius (°C) as follows:

$$°C = (°F - 32)/1.8$$

Vertical coordinate information is referenced to the National Geodetic Vertical Datum of 1929 (NGVD 29).

Horizontal coordinate information is referenced to the North American Datum of 1983 (NAD 83)

Elevation, as used in this report, refers to distance above the vertical datum.

*Transmissivity: The standard unit for transmissivity is cubic foot per day per square foot times foot of aquifer thickness [(ft^3/d)/ft^2]ft. In this report, the mathematically reduced form, foot squared per day (ft^2/d), is used for convenience.

Concentrations of chemical constituents in water are given either in milligrams per liter (mg/L) or micrograms per liter (µg/L).

Abbreviations

2D	two dimensional
BLS	below land surface
CCA	chromated copper arsenate
MASW	multichannel analysis of surface-wave
MCL	maximum contaminant level
NYC	New York City
NYSDEC	New York State Department of Environmental Conservation
NYCDEP	New York City Department of Environmental Protection
USGS	U.S. Geological Survey

Geophysical and Hydrologic Analysis of an Earthen Dam Site in Southern Westchester County, New York

By Anthony Chu, Frederick Stumm, Peter K. Joesten, and Michael L. Noll

Abstract

Ninety percent of the drinking water for New York City passes through the Hillview Reservoir facility in the City of Yonkers, Westchester County, New York. In the past, several seeps located downslope from the reservoir have flowed out from the side of the steepest slope at the southern end of the earthen embankment. One seep that has been flowing continuously was discovered during an inspection of the embankment in 1999. Efforts were made in 2001 to locate the potential sources of the continuous flowing seep. In 2005, the U.S. Geological Survey, in cooperation with the New York City Department of Environmental Protection, began a cooperative study to investigate the relevant hydrogeologic framework to characterize the local groundwater-flow system and to determine possible sources of the seeps. The two agencies used hydrologic and surface geophysical techniques to assess the earthen embankment of the Hillview Reservoir. Between April 1, 2005 and March 1, 2008, water levels were measured manually each month at 46 wells surrounding the reservoir, and flow was measured monthly at three of the five seeps on the embankment. Water levels were measured hourly in the East Basin of the reservoir, at 24 of 46 wells, and discharge was measured hourly at two of the five seeps. Slug tests were performed at 16 wells to determine the hydraulic conductivity of the geologic material surrounding the screened zone. Estimated hydraulic conductivities for 25 wells on the southern embankment ranged from 0.0063 to 1.2 feet per day and averaged 0.17 foot per day. The two-dimensional resistivity surveys indicate a subsurface mound of electrically conductive material (low-resistivity zone) beneath the terrace area (top of dam) surrounding the reservoir with a distinct elevation increase closer to the crest. Two-dimensional shear wave velocity surveys indicate a similar structure of the high shear wave velocity materials (high-velocity zone), increasing in elevation toward the crest and decreasing toward the reservoir and toward the northern part of the study area. Water-quality samples collected from 12 wells, downtake chamber 1 of the reservoir, and two seeps detected the presence of arsenic, toluene, and two trihalomethanes. Water-quality samples collected at the two seeps detected fluoride, indicating a connection with reservoir water.

Shallow wells on the southern embankment exhibited the largest seasonal water-level fluctuations ranging between 6 feet and 12 feet. The embankment is constructed from reworked low-permeability glacial deposits at the site. Water-level responses in observation wells within the embankment indicate that there is a shallow (approximately the upper 45 feet of the embankment) and a deep water-bearing unit within the embankment with a large downward vertical gradient between the shallow and deep water-bearing units. Precipitation strongly affected water levels in shallow wells, whereas the basin appears to be the main control on water levels in the deep wells. Seeps on the embankment slope appear to be caused by above-average precipitation that increases water levels in the shallow water-bearing unit, but does not easily recharge the deep water-bearing unit. Based on the data that have been analyzed, source water to the seeps appears to be primarily groundwater and, to a lesser extent, water from the East Basin of the reservoir.

Introduction

The Hillview Reservoir in Yonkers, southern Westchester County, New York, (fig. 1) was constructed between 1913 and 1916, contains more than 900 million gallons of water, and maintains a hydrostatic head of about 293 feet (ft) on the New York City (NYC) water-supply distribution system to the south. Ninety percent of NYC's drinking water passes through the Hillview Reservoir facility from the Kensico Reservoir, which is fed by the Delaware and Catskill aqueducts in upstate New York. Water is chlorinated at the reservoir and is piped from the southern end of the reservoir for distribution to users in NYC. The concrete lined reservoir, which has an area of about 90 acres, is about equally divided into the East Basin and West Basin by a concrete dividing wall, and has operated continuously since the first water tunnel was completed in 1917. A 14-ft-diameter concrete reinforced conduit (connecting conduit) hydraulically connects downtake chamber 1 to downtake chamber 2 (fig. 2; Malcolm Pirnie, Inc. and TAMS Consultants, Inc., 2002).

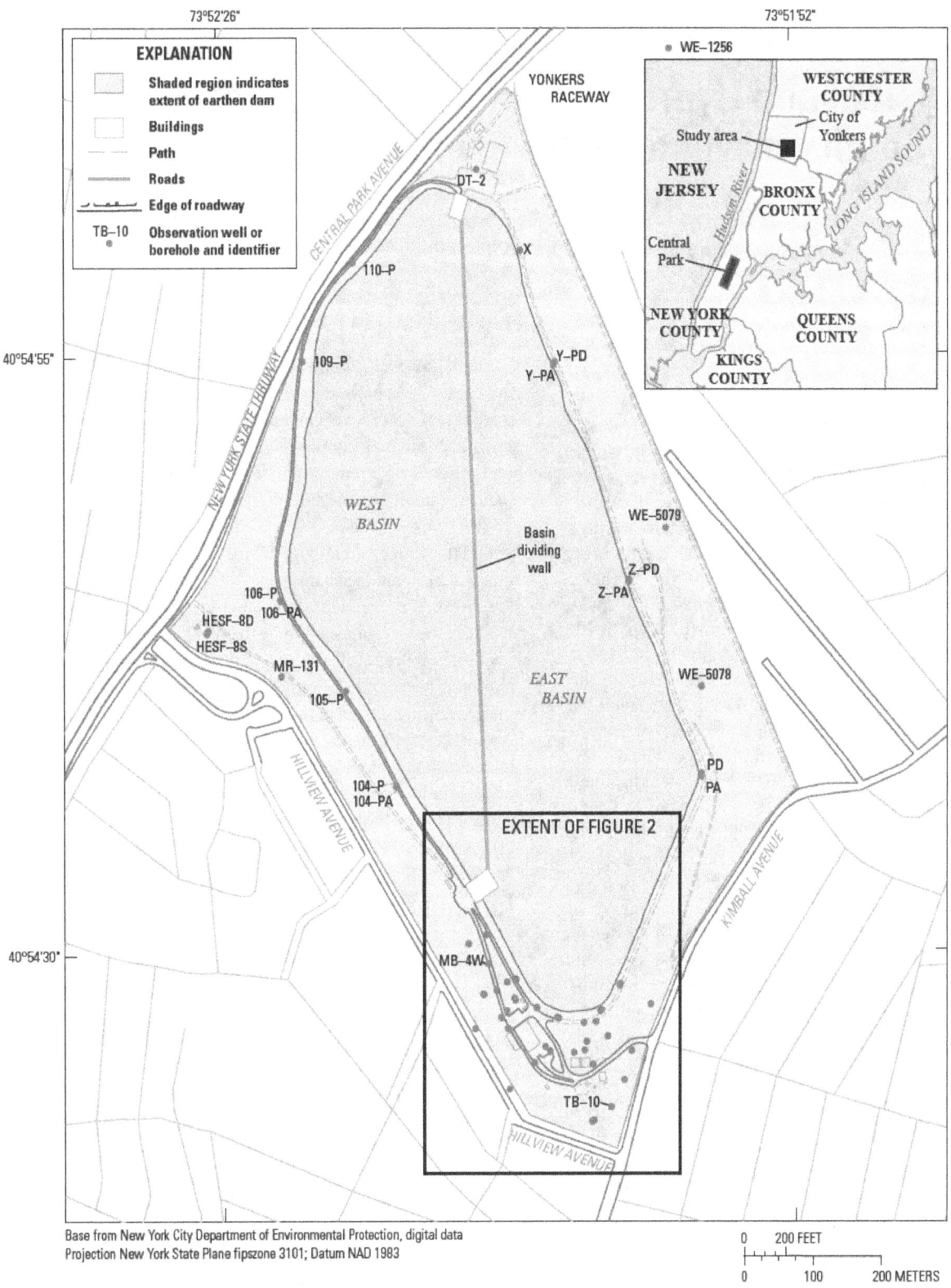

Figure 1.   Location of the Hillview Reservoir study area and selected observation wells and boreholes, City of Yonkers, Westchester County, New York.

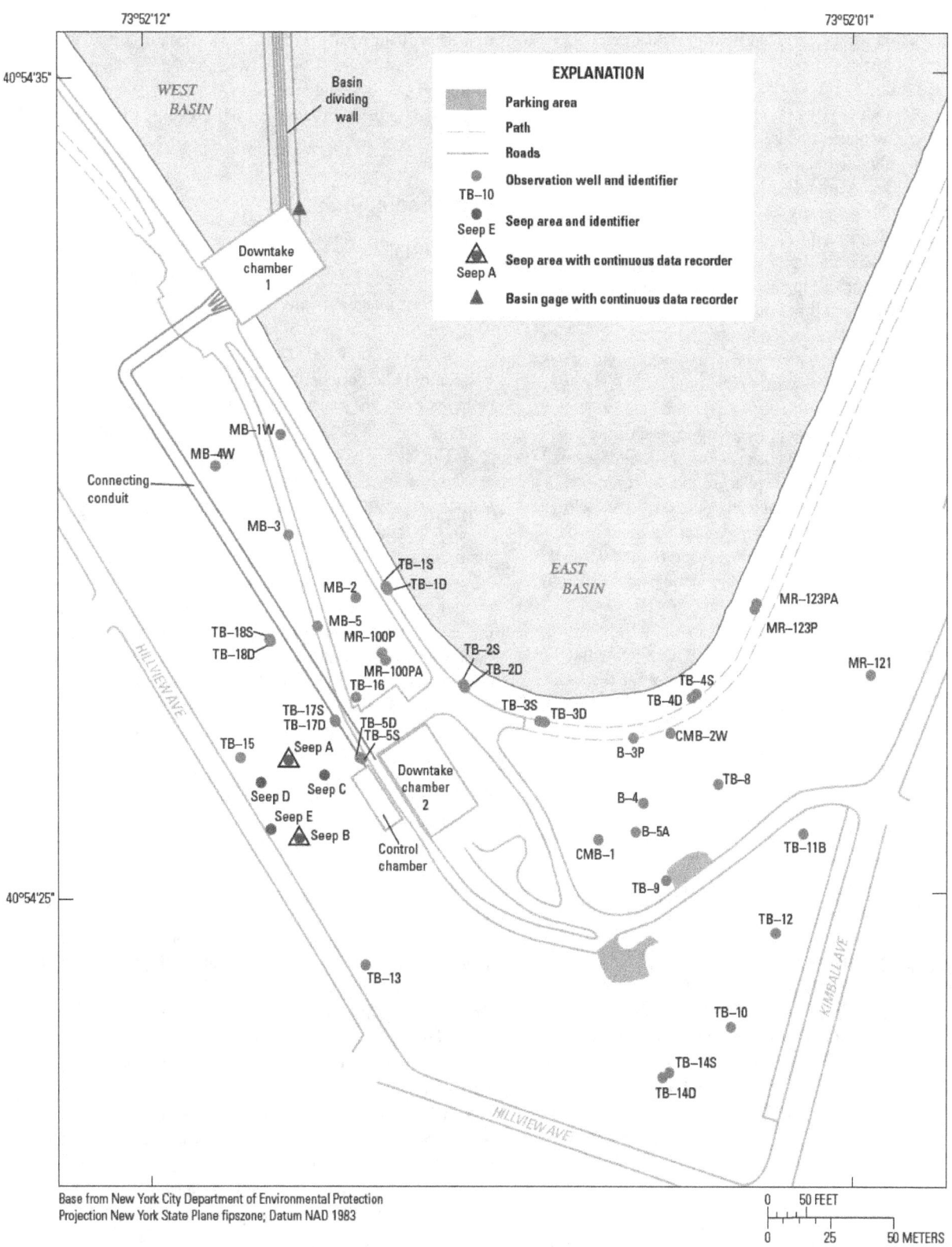

Figure 2.   Location of wells and seeps on the southern embankment at the Hillview Reservoir, City of Yonkers, Westchester County, New York.

The earthen embankment comprises low-permeability glacial clays that were excavated from the site and rest on a veneer of low-permeability glacial deposits that overlie crystalline bedrock. The earthen embankment was subsequently modified by other construction and maintenance projects near downtake, uptake, and control chambers; connecting shafts; connecting conduits; the reservoir dividing wall; and the bypass tunnel.

Groundwater generally flows outward from the reservoir toward the surrounding glacial drift. In the past, several seeps located downslope from the reservoir have flowed out from the side of the steepest slope at the southern end of the earthen embankment. One seep that flowed continuously during the study was discovered during an inspection of the embankment in 1999 (George Schmitt, New York City Department of Environmental Protection, written commun., 2007). In 2001, the New York City Department of Environmental Protection (NYCDEP) drilled 25 wells at the southern end of the reservoir, adding to the 32 wells previously installed around the reservoir (for a total of 57 wells around the reservoir) in an effort to locate the potential sources of the continuous flowing seep. The NYCDEP approach included taking periodic depth-to-water measurements and sampling reservoir and spring water for major ions. The results were inconclusive (Malcolm Pirnie, Inc. and TAMS Consultants, Inc., 2002).

In 2005, the U.S. Geological Survey (USGS) began a cooperative study with the NYCDEP to investigate the relevant hydrogeologic framework to characterize the local groundwater-flow system and to determine possible sources of the seeps. The USGS was able to access 46 of the 57 wells for monitoring water levels, temperature, and water quality. The principal objectives of this study were to (1) characterize the distribution of groundwater levels near the reservoir to obtain a better understanding of the embankment groundwater-flow system; (2) characterize possible flowpaths of water to the seeps; and (3) monitor the long- and short-term changes in groundwater levels and seep discharge within the study area.

Purpose and Scope

This report presents a summary of the hydrologic conditions in the southern part of the Hillview Reservoir, and hydrologic, surface geophysical, and basic water-quality data collected by the USGS to investigate the hydrogeologic framework of the southern embankment to determine the possible sources of seeps. Selected hydrographs of seasonal water-level elevations and distribution of hydraulic conductivities are shown in illustrations.

Study Area

The Hillview Reservoir in southern Westchester County in the City of Yonkers was put into service in 1917 when the first water tunnel was completed. The reservoir has a surface area of more than 90 acres and contains more than

900 million gallons of water (fig. 1). The reservoir is bounded to the north and west by the New York State Thruway, to the north and east by the Yonkers Raceway and residences, to the south and east by residences along Kimball Avenue, and to the south and west by residences along Hillview Avenue and a business district.

Hydrogeologic Setting

The Hillview Reservoir study area is underlain by unconsolidated Holocene deposits, artificial fill (modified glacial clays), and glacial-drift deposits of Pleistocene age. These sediments consist of boulders, gravel, sand, silt, and clay, which are underlain by crystalline bedrock. The bedrock is fractured and permeable where transmissive fractures or faults exist. In general, the bedrock forms a relatively impermeable base of the groundwater-flow system at the site.

The Hillview Reservoir earthen dam consists of an assemblage of reworked Pleistocene materials, manmade or artificial fill, and an underlying layer of Pleistocene glacial till and drift deposits. These unconsolidated deposits rest upon bedrock. The groundwater levels within the earthen embankment at the Hillview Reservoir are strongly affected by recharge from precipitation and from the reservoir to the unconsolidated embankment. The reservoir water levels fluctuate as a result of increased or decreased water demand during daily cycles. This cyclic demand produces an artificial "tidal" load on the surrounding embankment materials and local groundwater-flow system. References in this report to tidal effects on groundwater levels refer to this artificial tidal loading created by the rising and falling water levels within the reservoir basins.

Southern Westchester County is underlain by a high-grade metamorphic bedrock sequence consisting of gneiss, schistose-gneiss interlayered with granite, and marble (Asselstine and Grossman, 1955; Baskerville, 1982, 1992). The bedrock in southern Westchester County consists of a series of northeast-trending ridges and valleys. The ridges generally are underlain by gneiss and granite (Asselstine and Grossman, 1955; Baskerville, 1982). Hillview Reservoir is located on a ridge that is underlain by gneiss that is likely the Yonkers Gneiss or Fordham Gneiss. The bedrock contains many fractures, some of which are transmissive. The gneiss is considered a poor-to-moderate groundwater producer, whereas the marble is the most productive bedrock in Westchester County (Asselstine and Grossman, 1955). Depth to bedrock ranges from less than 1 ft below land surface (BLS) to 125 ft BLS within the southern part of Westchester County. The range in thickness of the natural till at the reservoir was estimated to be between 45 and 70 ft (Malcolm Pirnie, Inc., and TAMS Consultants, Inc., 2002); however, records of wells installed along the lower lying and northern areas of the reservoir indicate depths to bedrock are about 20 ft. Records for one well, WE–1256, to the north of the reservoir at Yonkers Raceway indicated a depth to bedrock of 24 ft (Asselstine and Grossman, 1955).

At least two groundwater-flow zones—one shallow and the other deep—appear to exist at the study area. Wells in the shallow flow zone have the highest water levels, are only slightly affected by reservoir tidal influences, and seem to respond to substantial precipitation recharge. In contrast, wells in the deep flow zone have lower water-level elevations, are highly affected by reservoir tidal influences, and only slightly respond to precipitation-induced recharge.

Methods of Investigation

The NYCDEP installed 57 wells in the earthen embankment surrounding Hillview Reservoir before this investigation. Only 46 of the original 57 wells were available for use in this study; the remaining 13 had been damaged during maintenance. At least five separate seeps located downslope from the reservoir have been documented at the Hillview Reservoir. The five seeps are identified as seeps A, B, C, D, and E. Seeps A, B, and E have been flowing the longest at the site (fig. 2).

Hydrologic data were collected from the 46 groundwater observation wells, a surface-water-level gage programmed to measure water-level and temperature fluctuations in the East Basin of the reservoir, and the four seeps that flowed during the study (table 1); seep D was not flowing during the time of the study. Surface-geophysical data were collected along the roadways and embankment slope in the vicinity of the seeps. Twelve wells, seeps A and E (which were flowing during the study), and the East Basin were sampled for water-quality analysis. Twenty-five wells were hydraulically tested to estimate the hydraulic conductivity of the earthen embankment in the southernmost part of the site.

Water-Quality Sample Collection

In winter and spring 2006, the USGS sampled 12 groundwater observation wells (TB–1S, TB–1D, TB–2D, TB–5S, TB–5D, TB–12, TB–15, TB–17S, TB–17D, TB–18S, TB–18D, and B–3P) to determine water-quality conditions in the westernmost part of the study area in relation to the seeps on the embankment. Samples were also collected from the East Basin (at downtake chamber 1) and at seeps A (from the flume) and E (along the stone wall) (fig. 2). Water samples were collected in accordance with standard USGS methods as described in the USGS National Field Manual for the Collection of Water Quality Data (U.S. Geological Survey, variously dated) and were analyzed at the USGS National Water Quality Laboratory (NWQL) in Denver, Colorado, for nutrients, major ions, metals, pesticides, pesticide degradates, and volatile organic compounds (table 2). Temperature, pH, and specific conductance were recorded in the field for each water sample at the time of collection and are listed in table 2.

Single Well Slug Test Methods

Water displacement tests, commonly known as "slug tests," were conducted in 16 wells at the Hillview Reservoir in August 2007 to estimate the hydraulic conductivity of the water-bearing zones in which the screened intervals of the wells were completed (table 3). The water in the well was displaced by a solid object, called a slug, and the water-level recovery was measured as a function of time. Based on the rate of recovery in the well, the hydraulic conductivity of the screened interval was calculated.

The slugs consisted of 1.07-inch-diameter (outside dimension) polyvinyl chloride pipes with caps on both ends. The slugs were suspended in the wells by a polypropylene rope attached to an eye bolt at the top of each slug, and the rope was secured to a section of pipe at the top of the well.

Water levels in the wells were measured during the slug tests using digital pressure transducers that were suspended in the water column by a cable that was secured at the top of the well. The pressure-transducer data loggers recorded water-level data based on the pressure of the water column in the well; water levels are accurate to 0.01 ft. The data loggers were programmed to record data on a logarithmic time scale. Water-level data were collected every 0.3 second at the beginning of the test, and the sampling interval between measurements was increased logarithmically to 5 minutes after 1 hour. This sampling interval provided detailed data coverage during the early portion of the slug test and less detailed coverage during the latter portion of the test, where less change was expected. The data loggers were started a few seconds before the insertion of the slug to record the background water level and to ensure that the exact time the slug was inserted was recorded. After the end of the test, the water-level data were downloaded to a computer and analyzed using slug test analysis software.

Care was taken to avoid dropping the slug in the water to minimize splashing because this could make early data from the test difficult to interpret. After the slug was inserted, the well was allowed to recover without any outside influence.

The slug tests ran for a minimum of 2 hours, and many of the tests ran for 5 hours or more. The duration of the tests was based on well construction, hydraulic characteristics, observed water-level changes under ambient conditions, and in some cases, previous hydrologic testing in the wells. The slug tests were allowed to run until the rate of recovery was less than 0.01 ft in a 5-minute interval.

Surface Geophysics

Surface-geophysical data were collected in the vicinity of seep A. The two surface-geophysical methods used were two-dimensional (2D) resistivity and 2D multichannel analysis of surface-wave (MASW) surveys.

The 2D resistivity surveys measure the subsurface electrical resistivity distribution by injecting electrical current

Table 1. Site information for A, groundwater wells and B, reservoir basin and seeps at the Hillview Reservoir, City of Yonkers, Westchester County, New York.

[All depths are in feet below land surface. Latitude and longitude are referenced to the North American Datum of 1983 (NAD 83). All elevations are in feet above National Geodetic Vertical Datum of 1929 (NGVD 29). NYSDEC, New York State Department of Environmental Conservation; ID, identifier; USGS, United States Geological Survey; —, no data]

A. Groundwater wells

Local well name	NYSDEC well ID	Longitude	Latitude	USGS site ID	Measuring point elevation	Land surface elevation	Well depth	Sounded depth	Screen depth	Saturated zone screened[1]	Sample date
TB-1S	WE5051	-735210	405428	405428073520901	302.01	300.69	40	41.8	30-40	Shallow	2/7/2006
TB-1D	WE5062	-735210	405428	405428073520902	302.37	300.49	122	72.05	60-70	Deep	2/7/2006
TB-2S	WE5058	-735206	405427	405428073520302	300.23	299.73	41	38.45	30-40	Shallow	2/7/2006
TB-2D	WE5072	-735209	405427	405427073520802	300.23	300.16	71	67.45	60-70	Deep	2/7/2006
TB-3S	WE5032	-735207	405427	405426073520701	300	300.28	41	39.5	30-40	Shallow	2/7/2006
TB-3D	WE5057	-735207	405427	405426073520702	300.2	300.29	71	62.2	60-70	Deep	—
TB-4S	WE5039	-735205	405427	405427073520401	302.79	300.29	—	43.3	—	Shallow	—
TB-4D	WE5045	-735205	405427	405427073520402	302.46	300.05	103	73.1	60-70	Deep	—
TB-5S	WE5024	-735210	405426	405426073521002	300.92	299.17	51	50.9	30-40	Shallow	5/1/2006
TB-5D	WE5071	-735210	405426	405426073521004	301.13	299.03	77	76.15	66-76	Deep	2/6/2006
TB-8	WE5040	-735204	405425	405425073524001	276.87	275.14	42	42.1	29-39	Shallow	—
TB-9	WE5043	-735204	405425	405424073520501	268.63	269.03	41	42.2	30-40	Toe	—
TB-10	WE5050	-735204	405423	405423073520401	245.44	243.39	41	42.35	20-40	Toe	—
TB-11B	WE5048	-735203	405425	405425073520201	255.85	253.46	31	31.7	20-30	Shallow	—
TB-12	WE5035	-735203	405427	405424073520301	249.27	247.11	52	32.1	20-30	Toe	5/1/2006
TB-13	WE5033	-735210	405424	405423073521001	220.66	217.97	31	32.4	10-30	Toe	—
TB-14S	WE5028	-735205	405422	405422073520501	241.45	238.9	48	—	20-30	Shallow	—
TB-14D	WE5041	-735205	405422	405422073520502	242.97	240.54	50	50.9	29-49	Toe	—
TB-15	WE5046	-735212	405426	405426073521201	229.77	227.95	33	30.17	12-32	Toe	5/2/2006
TB-16	WE5027	-735210	405427	405427073521001	299.33	299.75	49	48.75	—	—	—
TB-17S	WE5022	-735211	405427	405426073521001	299	297.32	40	39.9	30-40	Shallow	2/7/2006
TB-17D	WE5063	-735211	405427	405426073521003	299.2	297.37	80	79.75	70-80	Deep	2/7/2006
TB-18S	WE5056	-735212	405428	405427073521102	277.99	275.32	27	26.8	—	Shallow	5/2/2006
TB-18D	WE5049	-735212	405428	405427073521101	278.02	275.43	60	63.55	50-60	Deep	5/1/2006
MB-1W	WE5065	-735212	405430	405430073521101	300.38	300.55	107	77	60-80	—	—
MB-4W	WE5069	-735213	405430	405429073521201	288.31	286.82	103	61.5	50-60	Deep	—
MB-5	WE5070	-735211	405428	405427073521103	299.31	299.67	105	75.25	60-80	Deep	—
MR-100P	WE5042	-735210	405428	405427073520901	300.71	299.02	—	39.45	—	Shallow	—
MR-100PA	WE5066	-735208	405428	405427073520902	300.88	298.96	—	61.35	—	Deep	—

Table 1. Site information for A, groundwater wells and B, reservoir basin and seeps at the Hillview Reservoir, City of Yonkers, Westchester County, New York.—Continued

[All depths are in feet below land surface. Latitude and longitude are referenced to the North American Datum of 1983 (NAD 83). All elevations are in feet above National Geodetic Vertical Datum of 1929 (NGVD 29). ID, identifier; NYSDEC, New York State Department of Environmental Conservation; USGS, United States Geological Survey; —, no data]

A. Groundwater wells

Local well name	NYSDEC well ID	Longitude	Latitude	USGS site ID	Measuring point elevation	Land surface elevation	Well depth	Sounded depth	Screen depth	Saturated zone screened[1]	Sample date
MR–121	WE5067	-735202	405427	4054270735520102	265.19	262.33	—	20.5	—	Shallow	—
MR–123P	WE5068	-735204	405428	4054270735520801	302.49	299.73	—	—	—	—	—
MR–123PA	WE5031	-735204	405428	4054280735520301	302.06	299.35	—	—	—	—	—
MR–131	WE5074	-735222	405440	4054390735522101	273.2	270.6	33	32.55	—	—	—
B–3P	WE5026	-735206	405427	4054260735520501	299.79	300.53	—	22.55	—	Shallow	5/1/2006
B–4	WE5060	-735206	405426	4054250735520502	283.81	282.15	—	35.67	—	Shallow	—
B–5A	WE5055	-735206	405425	4054250735520501	279.45	277.35	—	35.45	—	Shallow	—
HESF–8S	WE5037	-735228	405443	4054420735522801	243.88	241.78	10	10.7	5–10	Shallow	—
HESF–8D	WE5029	-735228	405443	4054420735522701	244.15	241.81	19	19.3	14–19	Deep	—
CMB–2W	WE5064	-735205	405427	4054260735520502	301.03	299	—	35.4	—	Shallow	—
104–P	WE5059	-735217	405436	4054360735521702	301.72	300	—	41.3	—	—	—
104–PA	WE5047	-735217	405436	4054360735521701	302.98	300	—	20.7	—	—	—
105–P	WE5053	-735220	405441	4054400735522001	302.65	300.7	—	42.65	—	—	—
106–P	WE5061	-735224	405444	4054440735522301	301.2	298.9	—	42.4	—	—	—
106–PA	WE5054	-735224	405444	4054440735522401	301.33	296.6	—	42.62	—	—	—
109–P	WE5073	-735223	405454	4054540735522201	303.1	300.1	—	34.75	—	—	—
110–P	WE5034	-735220	405458	4054580735521901	303.4	302.1	—	32.65	—	—	—
111	—	-735215	405502	—	—	—	—	—	—	—	—
X	WE5021	-735209	405459	4054580735520901	302.1	299.9	—	38.35	—	—	—
Y–PA	WE5030	-735207	405454	4054540735520701	302.63	300	—	21.14	—	—	—
Y–PD	WE5044	-735207	405454	4054540735520702	302.45	300	—	42.4	—	—	—
Z–PA	WE5036	-735203	405445	4054450735520201	302.75	300.4	—	19.86	—	—	—
Z–PD	WE5052	-735203	405445	4054450735520202	302.06	299.4	—	46.12	—	—	—
PA	WE5038	-735159	405432	4054320735515801	303.17	300.3	—	22.05	—	—	—
PD	WE5023	-735159	405437	4054360735515801	302.76	300.2	—	42.3	—	—	—
DT–2	WE5025	-735212	405503	4055020735211101	300.9	298.92	—	44.8	—	—	—
WE–1256	WE1256	-735158	405508	4055080735515801	—	250	500	—	—	—	—
WE–5078	WE5078	-735157	405441	4054410735515701	290.6	287.6	124	124.2	—	—	—
WE–5079	WE5079	-735201	405449	4054490735520001	277.58	274.6	91	91.15	—	—	—

Table 1. Site information for A, groundwater wells and B, reservoir basin and seeps at the Hillview Reservoir, City of Yonkers, Westchester County, New York.—Continued

[All depths are in feet below land surface. Latitude and longitude are referenced to the North American Datum of 1983 (NAD 83). All elevations are in feet above National Geodetic Vertical Datum of 1929 (NGVD 29). ID, identifier; NYSDEC, New York State Department of Environmental Conservation; USGS, United States Geological Survey; —, no data]

B. Reservoir basin and seeps

Local site name	Longitude	Latitude	USGS site ID	Downstream order number	Land surface elevation	Sample date
East Basin	-735210	405433	405432073521001	1302006	—	2/8/2006
Seep A	-735210	405427	405426073521101	1302007	255.34	2/6/2006
Seep B	-735210	405426	405425073520901	1302009	234.14	—
Seep C	-735210	405426	405426073521005	—	259.84	—
Seep D	-735210	405426	405426073521006	—	234.61	—
Seep E	-735210	405426	405425073521101	1302008	221.61	2/8/2006

[1]The terms shallow and deep are used to identify wells screened in the shallow and the deep saturated zones. The term toe refers to wells located near the edge of the embankment and screened in a single water bearing unit where the shallow and deep saturated zones converge.

Table 3. Hydrologic properties of selected wells at the Hillview Reservoir in the City of Yonkers, Westchester County, New York.

[Well locations are shown in figure 2. —, no data]

Local well name	Delay in water-temperature response,[1] in months	Distance from reservoir,[2] in feet	Apparent linear groundwater velocity, in feet per month	Hydraulic conductivity, in feet per day
TB–1S	6	25	4.2	0.0097
TB–1D	5	25	5	0.0063
TB–2S	6	31	5.2	0.014
TB–2D	5	31	6.2	0.007
TB–3D	5	38	7.6	[3]0.351
TB–4S	6	31	5.2	[3]0.373
TB–4D	—	31	—	0.074
TB–5S	—	200	—	[3]0.076
TB–5D	—	200	—	0.0078
TB–8	—	169	—	0.023
TB–9	—	288	—	[3]0.02
TB–10	—	513	—	[3]0.07
TB–11B	—	300	—	[3]0.048
TB–12	—	400	—	0.01
TB–13	—	431	—	[3]0.096
TB–15	—	325	—	0.44
TB–17S	—	188	—	—
TB–17D	—	188	—	0.23
TB–18S	—	213	—	—
TB–18D	—	213	—	0.79
MB–1W	—	63	—	[3]0.017
MB–4W	18.5	150	8.1	1.2
MB–5	18	138	7.7	0.051
MR–100P	18.5	75	4	—
MR–100PA	19	75	4	—
B–3P	—	69	—	0.014
B–4	—	175	—	[3]0.294
B–5A	—	206	—	0.01
HESF–8D	—	396	—	—
CMB–2W	16	69	4	0.036
106–P	6	53	8.8	—
Z–PA	—	26	13	—
DT–2	17.5	132	7.5	—

[1] Delay between temperature change as recorded at the East Basin of the Hillview reservoir and corresponding temperature change in well.

[2] Linear distance from the subject well to the perimeter of the reservoir.

[3] Slug test data (unpublished) from TAMS Consultants, Inc. and analyzed by the U.S. Geological Survey.

into the ground through a pair of electrodes, then measuring the potential difference between a second pair of electrodes. The resistance of the subsurface material will be directly proportional to the measured potential difference and inversely proportional to the current injected (Zohdy and others, 1974). After a geometric correction is applied, the resultant resistivities are defined as apparent because they are based on a homogeneous subsurface.

Surface-wave surveys measure the shear, or surface wave, propagated from a seismic source. MASW, a recently developed seismic method, uses a seismic source in the form of a hammer or accelerated weight drop and an array of geophones that measures the ground roll (Rayleigh wave) generated from the nearby seismic source (Park and others, 1999). The multichannel recording method is effective in identifying and isolating noise from the data. The resulting one-dimensional shear wave data are inverted, and a 2D section is produced.

Hydrologic Measurements

Water-level elevations in all 46 wells were calculated from depth-to-water measurements collected with an electric water-level tape or a chalked steel tape. All the wells were monitored monthly throughout the course of the study. Of the wells in the southern embankment study area, 24 were instrumented with continuous-record digital pressure transducers programmed to measure water-level elevation and temperature hourly in each of the 24 wells. Volumetric measurements were taken monthly at seeps A, B, C, and E using a calibrated container and stopwatch. Continuous measurements were taken at seeps A and B with a digital pressure transducer programmed to measure stage and water temperature on an hourly schedule.

Water Quality

Water-quality samples were collected for this study from 12 groundwater observation wells, the East Basin, and at seeps A and E (fig. 2) to provide a basic understanding of the connections of the wells and the seeps to the basin. Pesticides and pesticide degradates were not detected in any of the samples collected. However, byproducts from disinfectants used in drinking water as well as metals and arsenic were detected in the samples collected. Arsenic concentrations greater than 1 microgram per liter (μg/L) were detected at 8 of the 15 sites sampled. The highest arsenic concentration (21.8 μg/L) was detected at well B–3P. This elevated concentration may have been affected by the proximity of a pressure-treated wood deck constructed at a nearby trailer. Chromated copper arsenate (CCA) is a chemical preservative used to pressure-treat wood to protect it from rotting due to insects and microbes. Studies indicate that residue from CCA can leach into soil (U.S. Consumer Product Safety

Commission, 2011). The sample collected from well B–3P also had the highest concentrations of metals and chloride of the 15 sites that were sampled. Elevated chromium and copper concentrations at well B–3P may also be attributed to the proximity of the pressure-treated wood deck.

Toluene was detected at concentrations greater than the detection limit at 5 of the 15 sites sampled. The concentrations at these five sites ranged from 4.78 μg/L at TB–2D, to 0.022 μg/L (estimated) at TB–17D (table 2).

Trihalomethanes are disinfection byproducts that are formed when chlorine is added to water for disinfection and reacts with naturally occurring organic matter or bromides. Two trihalomethanes—bromodichloromethane and trichloromethane—were detected in water samples collected from the study area. Trace amounts of bromodichloromethane were detected at downtake chamber 1, well TB–17S, and seeps A and E, with the concentration highest at the downtake chamber, then dropping off with distance from the downtake chamber where disinfection occurs. This drop off in concentration indicates a connection between the downtake chamber, well TB–17S, and seeps A and E, with the downtake chamber as the source of the tracer compound (tables 1 and 3). Trichloromethane was detected at 13 of the 15 sites sampled. With the exception of the TB–17 paired wells (pair of shallow and deep wells), the deeper wells had the highest trichloromethane concentrations. Trichloromethane, an immiscible dense nonaqueous phase liquid that sinks when mixed in water, was not detected in samples collected from shallow wells TB–1S, B–3P, and TB–18S. Concentrations of trichloromethane in samples collected at well TB–15, downtake chamber 1 (East Basin), and seeps A and E were in the middle of the range of concentrations for the deep and shallow well samples, which suggests that seeps A and E may have a connection with the basin.

Municipalities add flouride to their drinking water to support dental health. Flouride, which is added to the New York City water supply, was detected in samples collected from both seeps at concentrations of 0.82 milligrams per liter (mg/L) at seep A, and 0.77 mg/L at seep E, indicating a connection with water from the reservoir (table 2).

Hydrogeology

All the wells monitored in this study were registered with the New York State Department of Environmental Conservation, and New York State well identification numbers were obtained for all the wells (table 1). Digital pressure transducers were installed in the East Basin (as a basin gage), at selected observation wells, and at two seeps to monitor hourly, daily, and seasonal changes in groundwater levels and temperature in relation to East Basin water levels and precipitation. Groundwater levels in all observation wells, discharge amounts at flowing seeps, and water levels at the East and West Basins were measured manually at monthly and quarterly intervals. Precipitation amounts were recorded by the National Weather Service at Central Park, New York, and were compared with the 138-year record. Surface geophysical methods were used to delineate variations in the subsurface geology.

Reservoir Tidal Effects

The Hillview Reservoir consists of two separate basins known as the East and West Basins. Both basins store water from several aqueducts and are used to maintain a constant hydraulic head within NYC's water supply system. A digital pressure transducer was installed in the East Basin. Water levels and temperature were recorded at 1-hour intervals. The data indicate that the water level in the basin ranges by 3 to 4 ft several times a day, creating an artificial basin tidal cycle (fig. 3). The effect of the basin tide on the water levels in the wells on the southern embankment appears to be related to the distance of the wells from the East Basin and the depth of the wells. Analysis of the hydrographs (fig. 3) indicates that water levels in the wells that are closest to the East Basin (fig. 2) and screened in the deep saturated zone are most strongly affected by the basin tidal cycle. This basin tidal cycle is referred to in this report as tidal or the basin tide. The mean water elevation within the East Basin is 293 ft (fig. 3).

Precipitation and Groundwater Levels

The National Weather Service defines (meteorological) spring as March through May, summer as June through August, autumn as September through November, and winter as December through February. The National Weather Service has maintained a precipitation recording station in Central Park since 1869. The 138-year mean annual precipitation is 44.98 inches. Based on the precipitation records, it appears that the region is in an extended wet period that began in 2002. The 138-year annual average has been exceeded every year from 2002 to 2007, and 7 of the previous 8 years dating back to 2000 (fig. 4). In addition, since 2003, total annual precipitation has exceeded 50 inches. Only once since 1970 has annual precipitation exceeded 50 inches for more than 2 years in a row. In 2003, 2004, 2005, 2006, and 2007, the annual precipitation exceeded the 138-year average by 30, 16, 24, 33, and 37 percent, respectively. According to the National Oceanic and Atmospheric Administration (2008) data from Central Park, 2006 and 2007 were the seventh and fourth wettest years recorded, respectively. October 2005, April 2007, September 2004, September 2005, June 2006, November 2006, and February 2008 were the first, second, third, fourth, fifth, eighth, and ninth wettest months ever, respectively. Conversely, September 2005 was the fourth driest month recorded. Seasonally, autumn 2005 and summer 2006, spring 2007 and summer 2007, and autumn 2006 were the third (tied), fourth (tied), and tenth wettest of those seasons ever, respectively. April 15, 2007, is the date of the second greatest daily precipitation ever recorded (7.57 inches) at Central Park.

All the wells showed a downward trend in water-level elevations during spring 2006. This trend corresponds to a relatively dry period that occurred that spring, which included the driest March in the Central Park precipitation record. Water levels appeared to rebound in response to increased precipitation in the months that followed the dry period. Most of the hydrographs presented in appendix 1 show an upward trend in water levels during that period. During April 2007, the hydrographs of all the instrumented wells documented a sharp upward shift in water-level elevation. This abrupt shift correlates with the largest single precipitation event during this study, where 7.57 inches of precipitation fell at Central Park within a 24-hour period.

Slug Test Analysis

The slug test data were analyzed using the Bouwer and Rice (1976) method with spreadsheets developed by Halford and Kuniansky (2002) using Microsoft Excel. The ratio of the change in water level to the initial change in water level after the insertion of the slug was plotted log-linearly as a function of time. A line was fit to the data points. Using the slope of this line and the well construction and screen length, hydraulic conductivity was calculated (Bouwer and Rice, 1976).

One common problem with slug test analyses is that the data often are not log-linear for the duration of the test. One reason these data are not log-linear is that well-screen intervals commonly are surrounded by sand packs, which often have a higher hydraulic conductivity than native aquifer materials. The groundwater stored in the sand pack quickly enters the well, whereas water from the aquifer enters the well at a much slower rate. Typically a dataset collected from a well with a substantial sand pack will have two log-linear sections of the dataset—a steep-dipping trend in early time from the water in the sand pack and a shallow-dipping trend in later time from the aquifer.

The analysis of the data collected at Hillview Reservoir was complicated by variations in well construction. Different screen lengths, drilled apertures, sand-pack properties, and missing or inaccurate records made comparisons of the interpretations difficult. At some of the wells in the southwestern area of the reservoir, the water level recovered too quickly to determine whether the measured hydraulic conductivity was from the sand pack or from the water-bearing unit. Consultants for the NYCDEP had previously completed slug tests on many wells (Malcolm Pirnie, Inc. and TAMS Consultants, Inc., 2002), but some of these wells were subsequently destroyed before the USGS study began. Data from these previous slug tests were reanalyzed by the USGS using the methods described above. Data from 9 consultant slug tests and 16 USGS slug tests (a total of 25 wells) were analyzed to estimate hydraulic conductivity (table 3).

Estimated hydraulic conductivities range from 0.0063 to 1.2 feet per day (ft/d) and averaged 0.17 ft/d (table 3). The contoured data shown in figure 5 indicate that an area

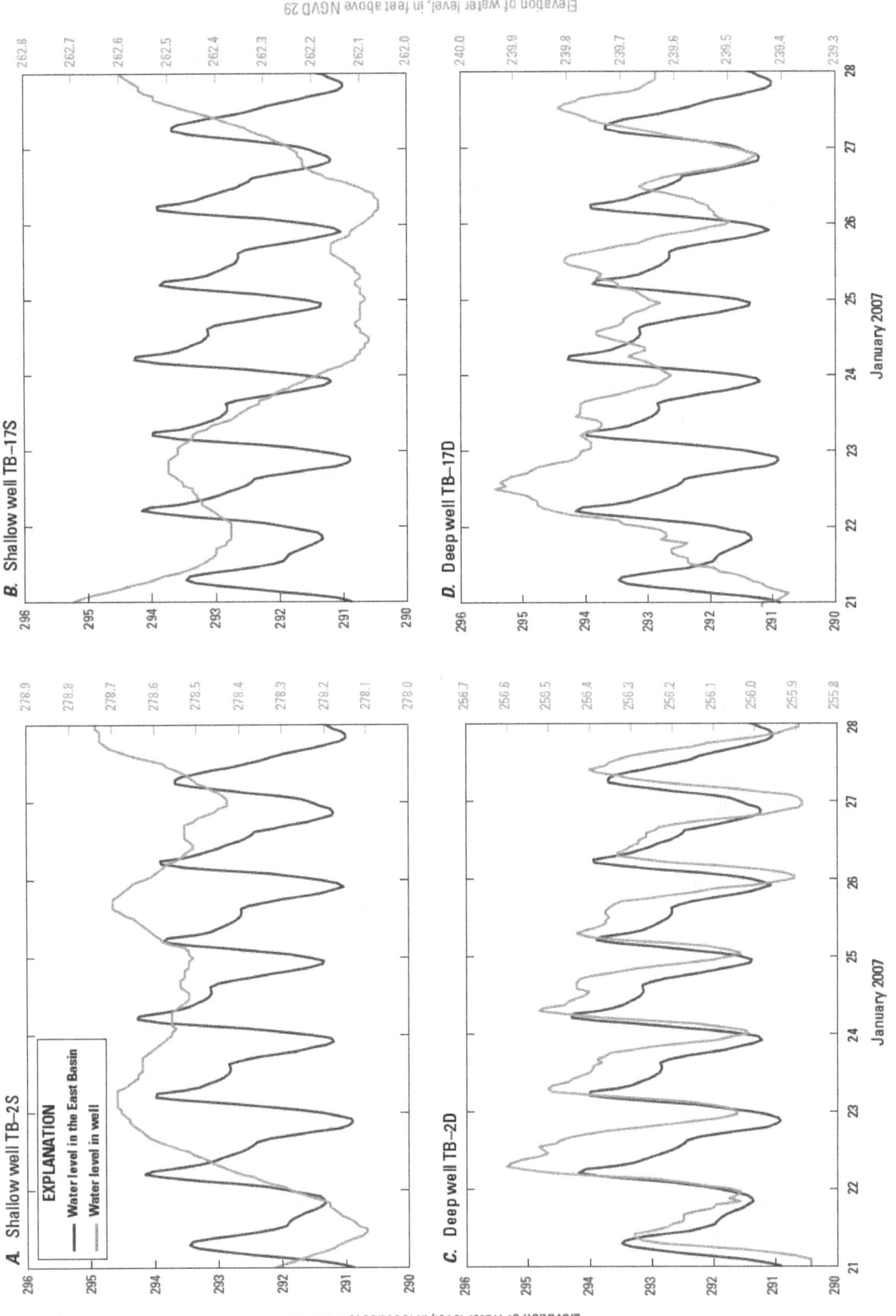

Figure 3. Response of water levels in nearby wells A, TB–2S; B, TB–17S; C, TB–2D; and D, TB–17D to East Basin water levels in the Hillview Reservoir, City of Yonkers, Westchester County, New York, in 2007. (NGVD 29, National Geodetic Vertical Datum of 1929.)

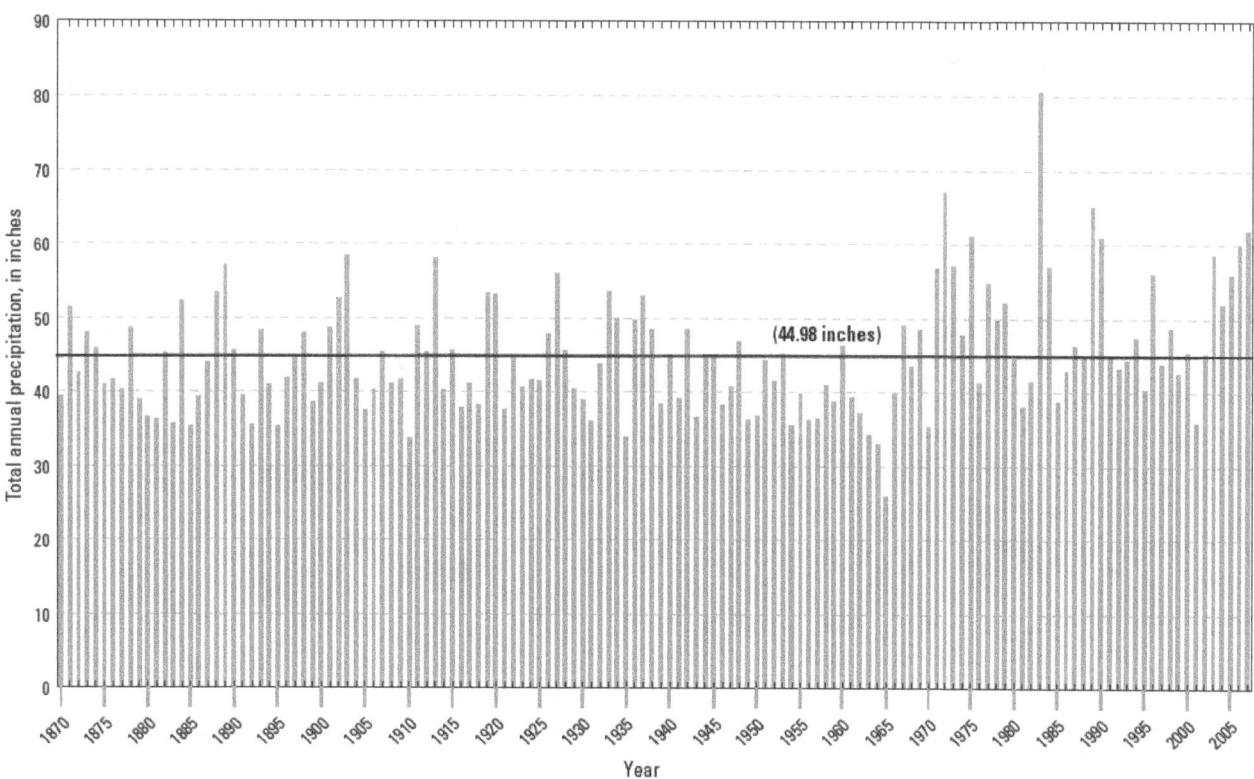

Figure 4. Annual precipitation as recorded at Central Park, New York County, New York, during 1870–2007. Solid black line indicates annual mean precipitation during the 138-year record of observation. Data are from National Oceanic and Atmospheric Administration (2008); modified from Stumm and others (2012).

of lower than average hydraulic conductivity appears to be present between the TB–17 and TB–3 well pairs and south toward Hillview Avenue along the southwestern part of the earthen dam. The hydraulic conductivity appears to increase away from this area toward the northwestern part and to some degree the northeastern part of the study area. There appears to be an abrupt transition from low permeable units to higher permeable units between wells MB–1W and MB–4W in the northwestern part of the study area (fig. 5). Two isolated areas of higher than average permeable deposits near wells TB–3D, B–4, and TB–4S were also indicated (figs. 2 and 5).

The lowest hydraulic conductivities in the shallow water-bearing unit were measured in wells TB–1S, B–5A, B–3P, and TB–2S at 0.0097 ft/d, 0.01 ft/d, 0.014 ft/d, and 0.014 ft/d, respectively (figs. 2 and 5; table 3), whereas the lowest hydraulic conductivities in the deep water-bearing unit were measured in TB–1D, TB–2D, TB–5D, and MB–5 at 0.0063 ft/d, 0.007 ft/d, 0.0078 ft/d, and 0.051 ft/d, respectively (figs. 2 and 5; table 3). The highest hydraulic conductivities in the shallow water-bearing unit were measured in wells TB–4S, B–4, CMB–2W, and TB–8 at 0.373 ft/d, 0.294 ft/d, 0.036 ft/d, and 0.023 ft/d, respectively (figs. 2 and 5; table 3), whereas the highest hydraulic conductivities in the deep water-bearing unit were measured in wells MB–4W, TB–18D,

TB–3D, and TB–17D at 1.2 ft/d, 0.79 ft/d, 0.351 ft/d, and 0.23 ft/d, respectively (figs. 2 and 5; table 3). Due to the variation in well construction and scarcity of data, no inference with respect to hydraulic conductivity and depth could be made with the data available.

Shear Wave Velocity and Resistivity Survey Observations

The 2D resistivity surveys indicate a subsurface mound of electrically conductive material (low-resistivity zone) beneath the terrace area (top of dam) surrounding the reservoir with a distinct elevation increase closer to the crest. The conductive zone appears to decrease in elevation toward the reservoir beneath the terrace (A–A′ and C–C′) and downslope toward Hillview Avenue (B–B′, D–D′, and E–E′) (figs. 6 and 7A–E).

Two-dimensional shear wave velocity surveys indicate a similar structure of the high shear wave velocity materials (high-velocity zone), increasing in elevation toward the crest and decreasing toward the reservoir and toward the northern part of the study area. The bedrock was imaged along cross-section F–F′ where depths ranged from 35 meters (m)

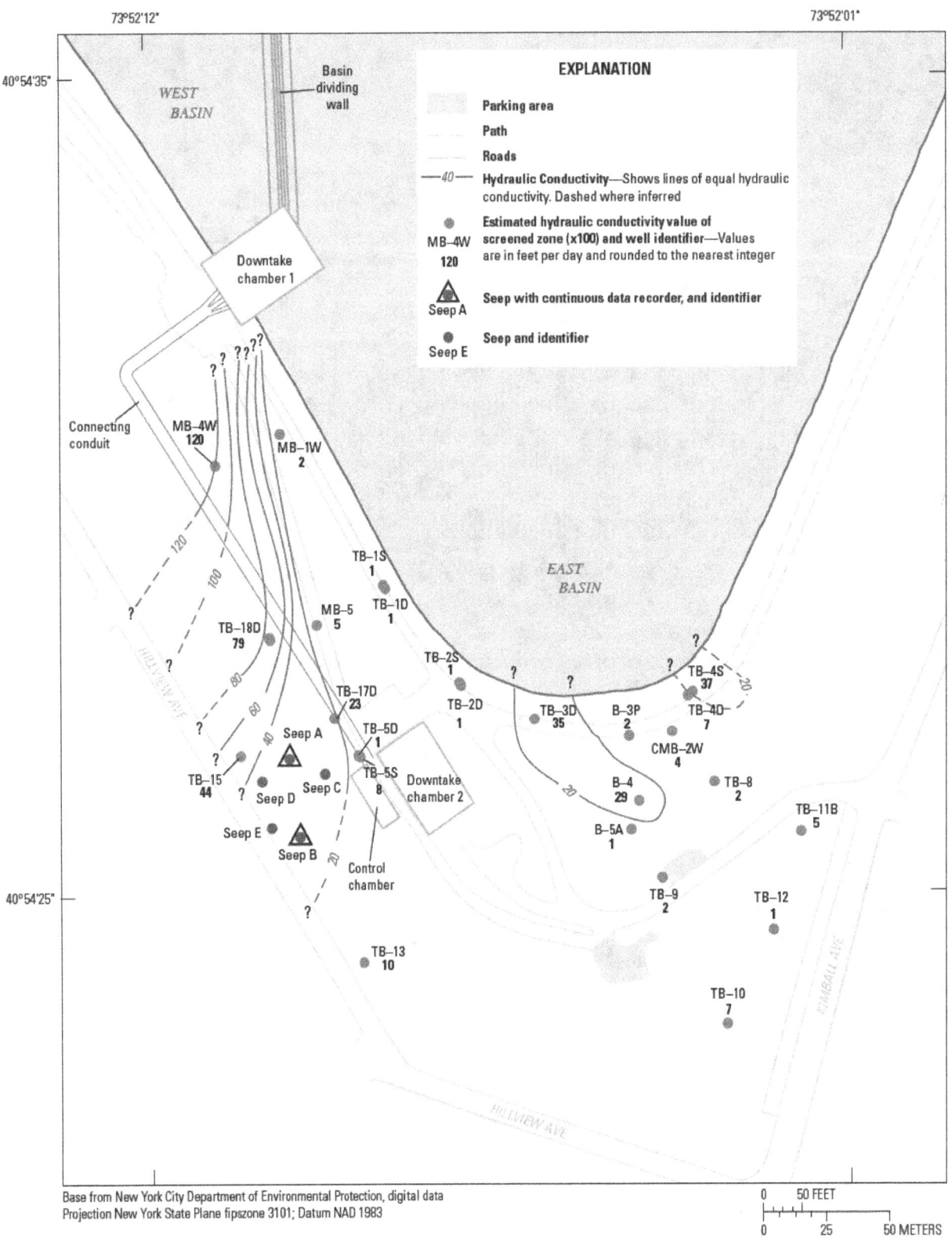

Figure 5.   Distribution of hydraulic conductivity estimated from slug test results of wells on the southern embankment at Hillview Reservoir, City of Yonkers, Westchester County, New York. Hydraulic conductivity values are listed in table 3 and are shown in this figure multiplied by 100 and rounded to the nearest integer.

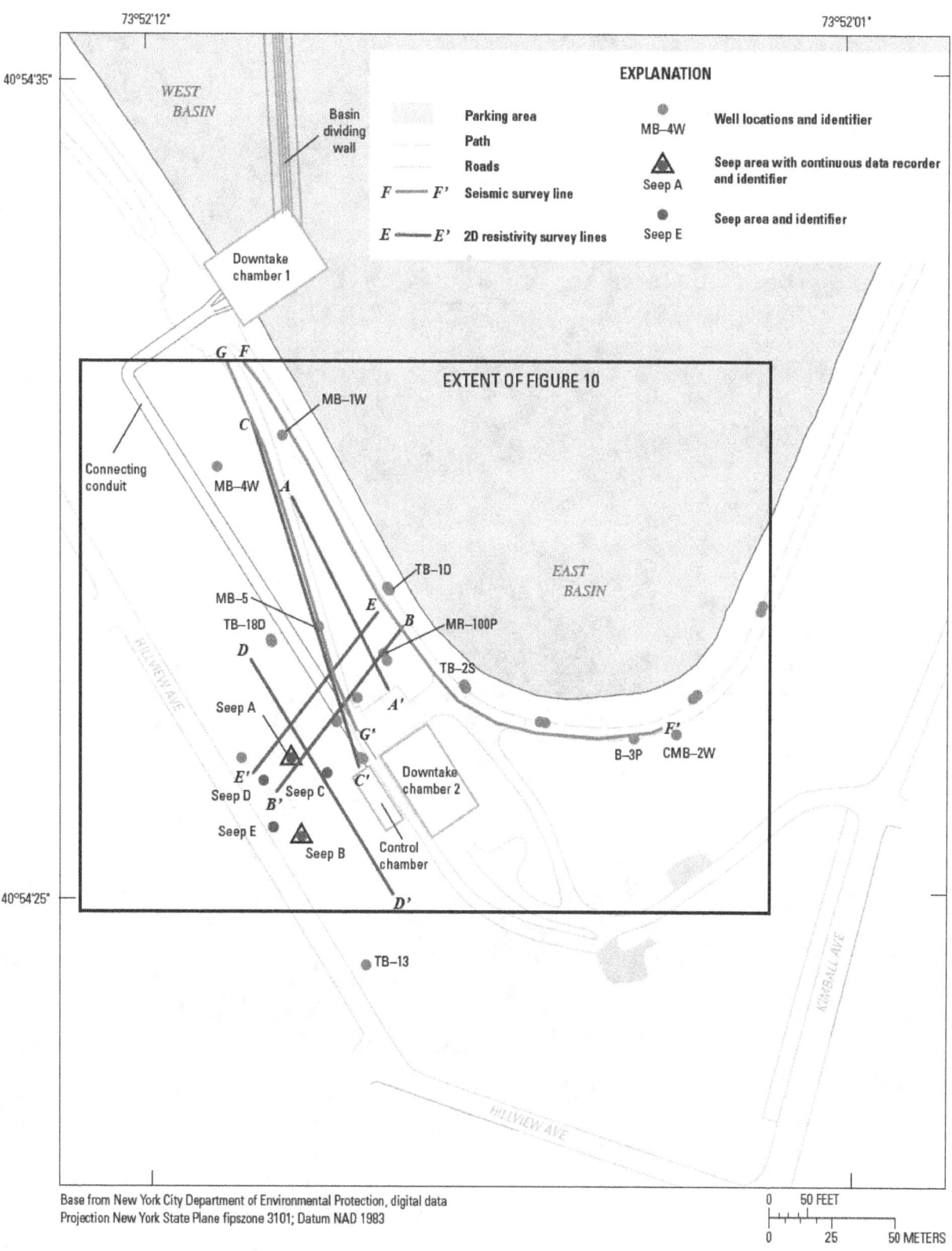

Figure 6. Locations of geophysical surveys on the southern embankment at the Hillview Reservoir, City of Yonkers, Westchester County, New York. 2D, two dimensional.

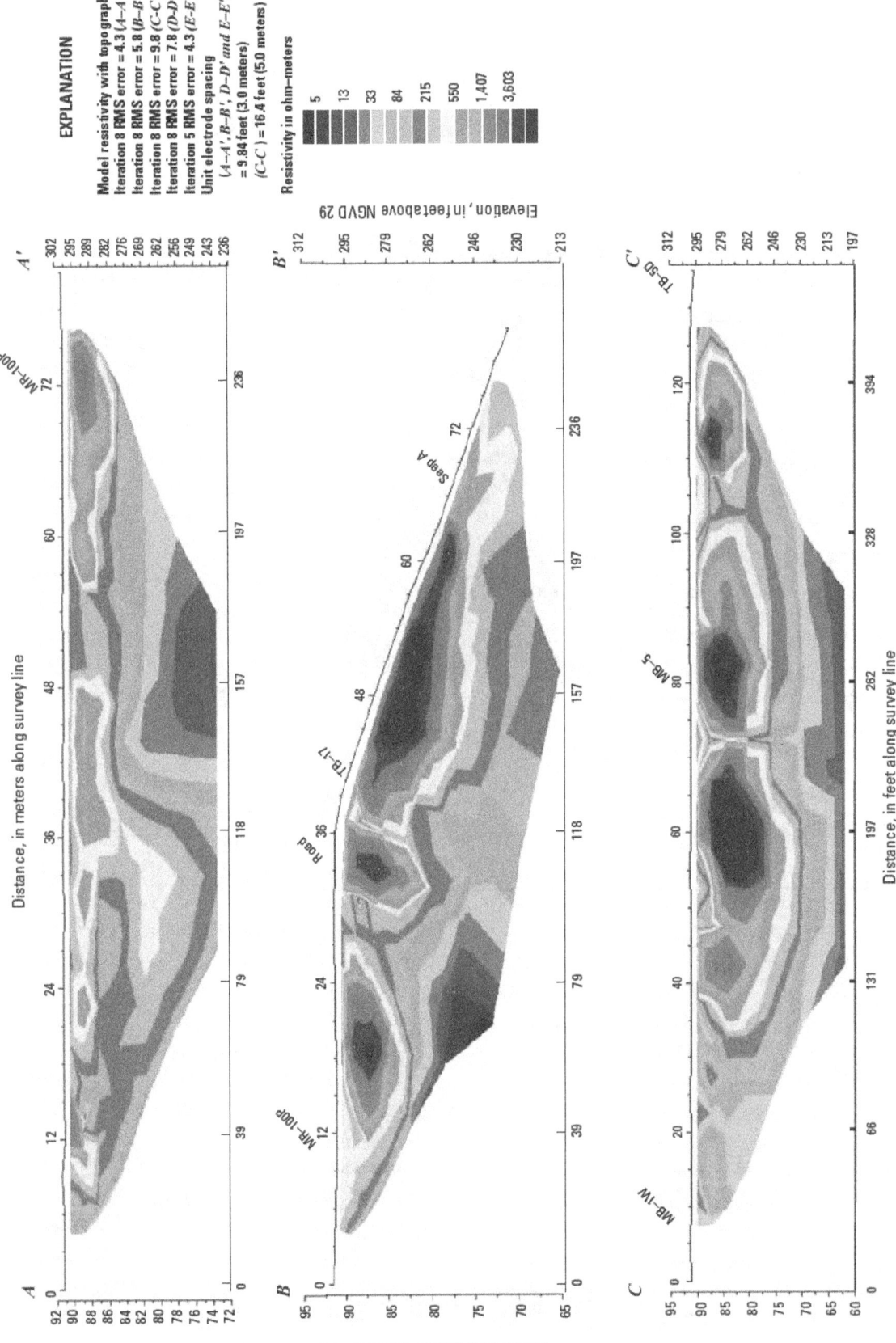

Figure 7. Cross-sections showing two-dimensional resistivity along A, A–A′; B, B–B′; C, C–C′; D, D–D′, and E, E–E′ at the Hillview Reservoir, City of Yonkers, Westchester County, New York. The locations of the cross-sections are shown in figure 6. Profiling was through a Schlumberger array. ft, foot; m, meter; RMS, root mean square.

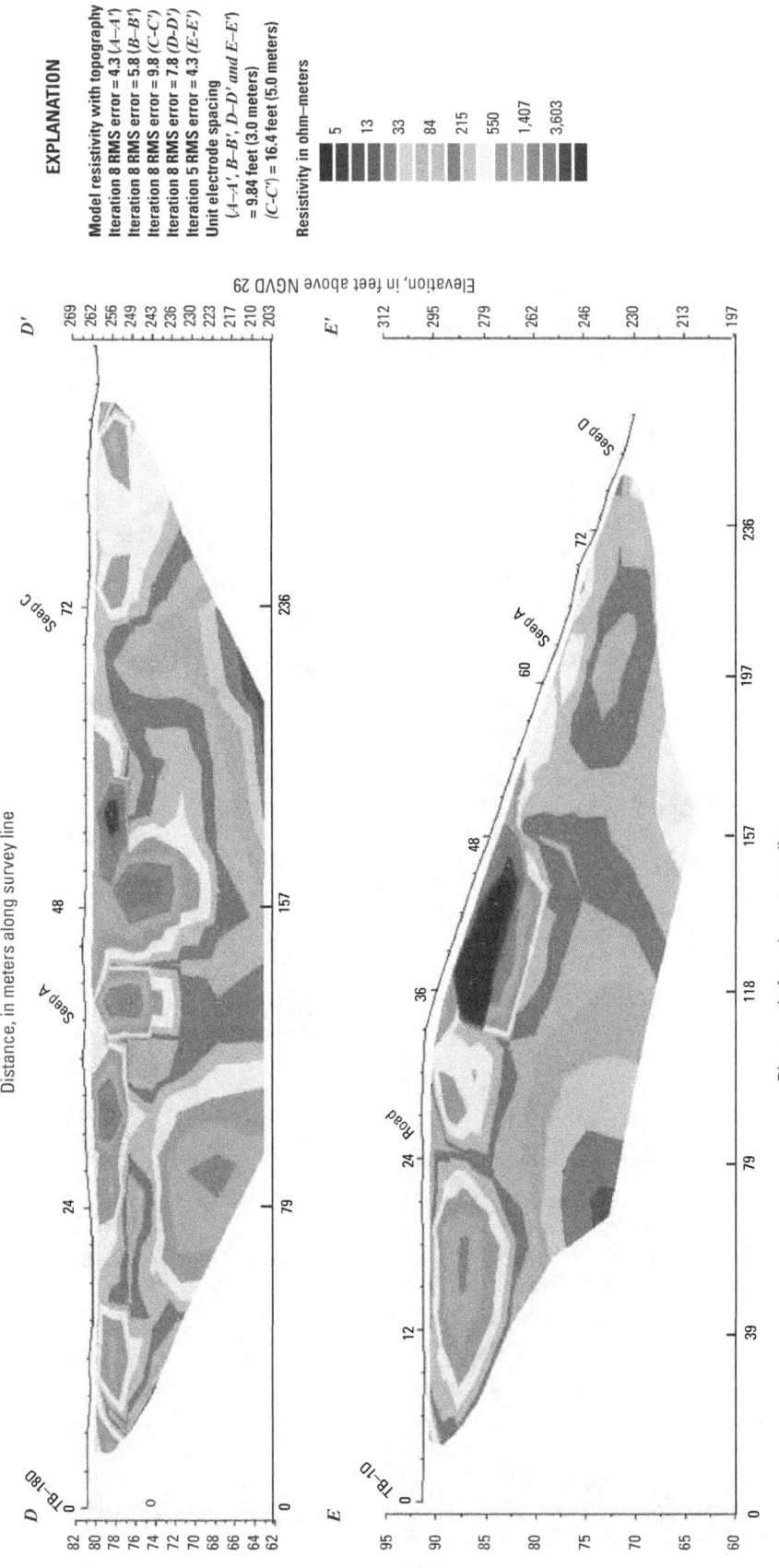

Figure 7. Cross-sections showing two-dimensional resistivity along A, A–A′, B, B–B′, C, C–C′, D, D–D′, and E, E–E′ at the Hillview Reservoir, City of Yonkers, Westchester County, New York. The locations of the cross-sections are shown in figure 6. Profiling was through a Schlumberger array. ft, foot; m, meter; RMS, root mean square.—Continued

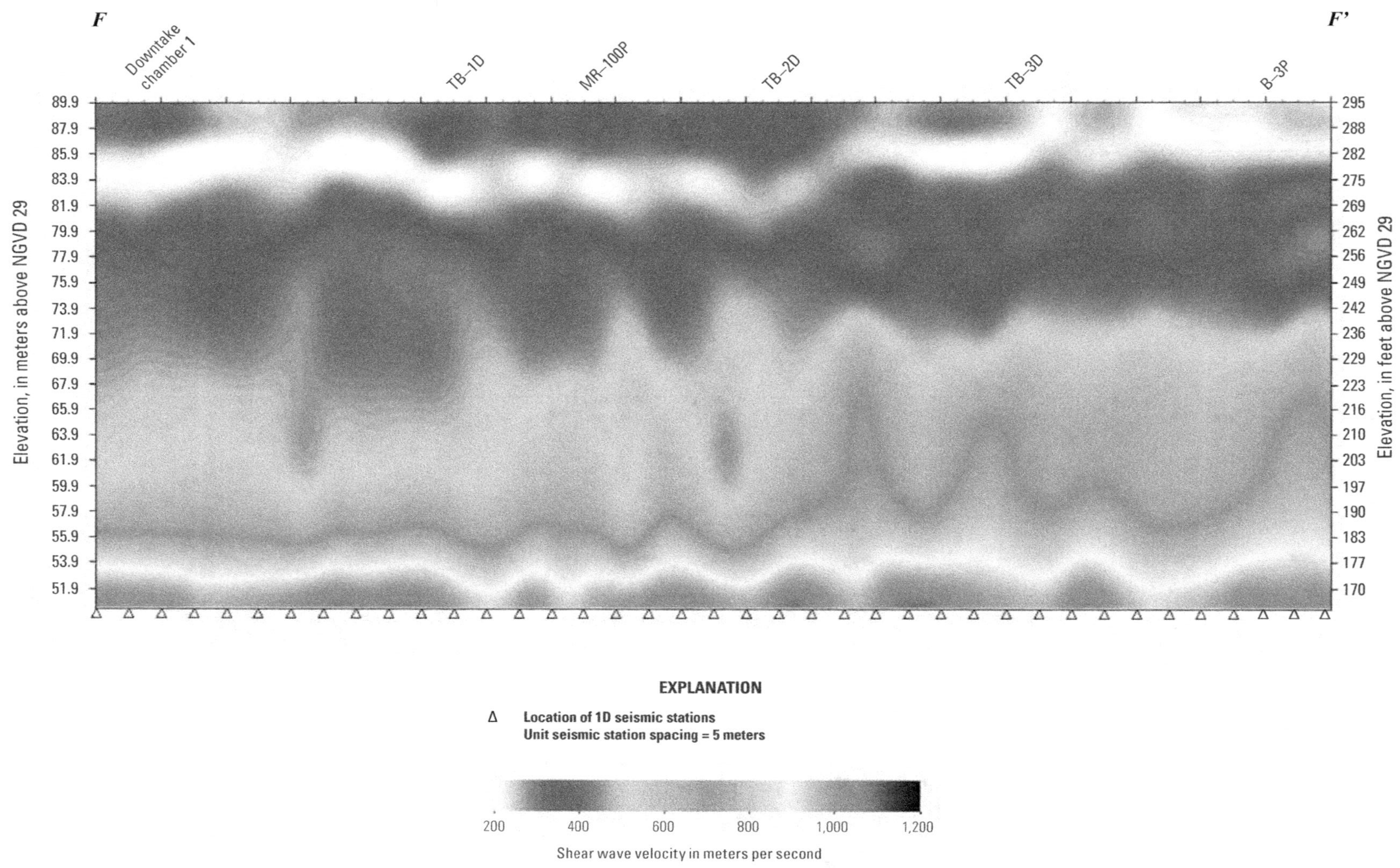

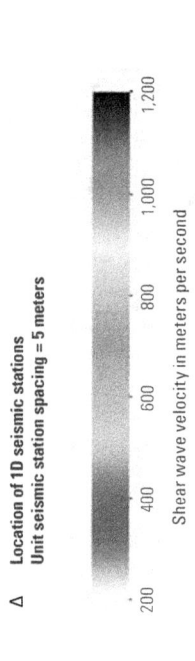

Figure 8. Cross-section showing two-dimensional shear wave velocity along F–F′ at the Hillview Reservoir, City of Yonkers, Westchester County, New York. The location of the cross-section is shown in figure 6. 1D, one-dimensional; NGVD 29, National Geodetic Vertical Datum of 1929.

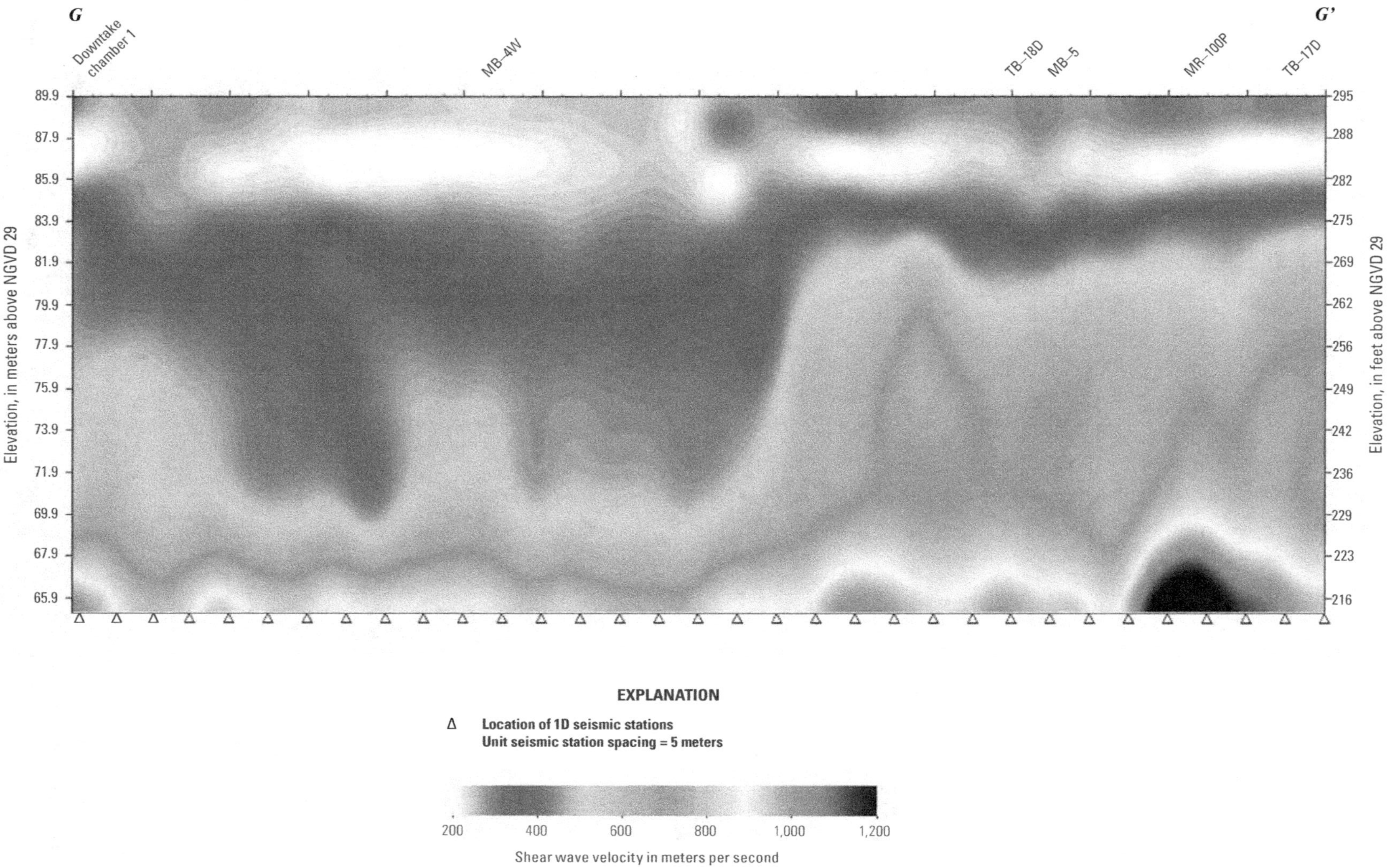

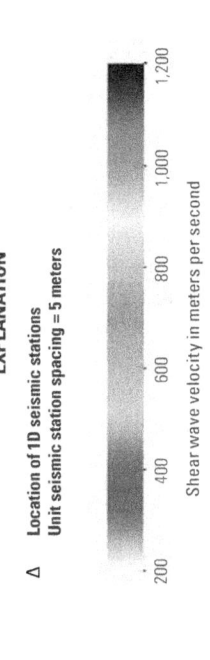

Figure 9. Cross-section showing two-dimensional shear wave velocity along G–G' at the Hillview Reservoir, City of Yonkers, Westchester County, New York. The location of the cross-section is shown in figure 6. 1D, one-dimensional; NGVD 29, National Geodetic Vertical Datum of 1929.

(114.8 ft) BLS near B–3P to 36 m (118.1 ft) BLS at TB–1D (fig. 8). The depth of the imaged bedrock along cross-section G–G′ ranged from 20 m (65.6 ft) to greater than 24 m (78.7 ft). A high-velocity zone that decreases in elevation from south to north along G–G′ (fig. 9) correlates with the decrease in elevation of conductive material from south to north along the 2D resistivity cross-section A–A′ (fig. 7).

Velocities in the high-velocity zone were on average almost 200 meters per second (656.2 feet per second) greater than the overlying material (figs. 8 and 9). This high-velocity zone ranges from only 6 m (19.7 ft) BLS near well MB–5 to almost 28 m (91.9 ft) BLS north of the TB–1 paired wells (figs. 8 and 9). When the low-resistivity zone (fig. 7) and the high-velocity zone (figs. 8 and 9) are compared, they seem to correlate within a few meters. This finding indicates that a zone of high electrical conductivity and high shear wave velocity underlies the westernmost part of the study area. This zone may be a well consolidated unit with higher clay content than the surrounding material. The water table appears to have less of an effect on resistivity in the most conductive material.

A contour map of the top of the apparent clay-rich zone was constructed, which combined both the 2D resistivity (low-resistivity zone) and shear wave velocity survey elevations (high-velocity zone) (fig. 10). This unit of low resistivity (clay-rich) and high shear wave velocity (dense) appears to be at its highest elevation in the vicinity of wells TB–16, MR–100P, MR–100PA, and MB–5 (fig. 10). The unit ranges in elevation from a maximum of 86 m (282.2 ft) above NGVD 29 near well MR–100P to a minimum of 64 m (210 ft) above NGVD 29 north of well TB–1D (fig. 10). Three areas were less conductive than other areas along the slope below wells TB–17 and TB–5 and may indicate gaps in the dense clay-rich material (figs. 5 and 10).

The unsaturated zone shows moderate-resistivity zones that correlate with low-velocity horizons at 2 to 8 m (6.6–26.2 ft) BLS, which may correlate to a sandier horizon that has less artificial compaction. Conductive (low-resistivity) zones were present in the deep subsurface. The elevation and distribution of these low-resistivity zones correlate with a high-velocity zone detected in the deep subsurface along both seismic survey lines F–F′ and G–G′ (figs. 6, 8, and 9).

Hydrologic Conditions

Groundwater levels were used to define the shallow and deep groundwater zones at the Hillview Reservoir. In addition, groundwater levels were evaluated to determine the hydrologic conditions in the earthen embankment.

Groundwater Levels

Groundwater levels were measured manually at 46 observation wells during synoptic surveys and recorded digitally at selected wells at hourly intervals. Evaluation of groundwater levels at the Hillview Reservoir site indicates the presence of at least two separate saturated zones—a shallow water-bearing unit and a deep water-bearing unit. The shallow saturated zone had the highest water levels and appears to be most affected by precipitation. Groundwater levels in the wells closest to the reservoir were slightly affected by basin tides; the shallow wells farther away were not affected by the artificial tides created by the fluctuations in the water levels of the reservoir's East Basin. During the study, shallow wells exhibited the largest seasonal water-level fluctuations, ranging between 6 ft and 12 ft at wells MR–100P and TB–17S, respectively (fig. 11). Within the south embankment study area, water levels in the shallow saturated zone fluctuated an average of 7 ft for all wells during this investigation. The largest fluctuations of 19 ft, 12 ft, and 10 ft occurred at wells TB–18S, TB–17S, and TB–13, respectively (appendix 1).

During the study period from April 2005 through February 2008, variation of water levels within 100 ft of the reservoir ranged from 3 to 6 ft in the shallow saturated zone and from 4 to 5 ft in the deep saturated zone. During this same period, variation of water levels in wells beyond 100 ft of the reservoir in the shallow saturated zone ranged from 5 to 20 ft whereas water levels in the deep saturated zone ranged only 2 to 5 ft. Observation wells TB–10, TB–12, and TB–14D, which are located southeast along the toe or edge of the embankment, had the largest range in water levels (12 to 20 ft) (appendix 1). In contrast, observation wells TB–13 and TB–15 had a relatively small variation with only a 2- to 4-ft range in water levels during the study, which indicates that the hydrologic characteristics of the embankment materials at wells TB–10, TB–12, TB–13, TB–14D, and TB–15 are highly variable.

Shallow Saturated Zone

The shallow water-bearing unit at the southern embankment of the Hillview Reservoir was defined as the uppermost 45 ft of the embankment materials below the crest of the reservoir. Most of the materials within the shallow water-bearing unit consist of modified glacial till and artificial fill. The material in the shallow water-bearing unit appears to be variable in composition with respect to the amount of fine-grained sediments. Hydrographs of observation wells TB–2S and TB–17S, screened in the shallow water-bearing unit, indicate water levels near the reservoir (well TB–2S) were only slightly tidally affected, and water levels about 150 ft from the reservoir (well TB–17S) were nontidal (fig. 3). Average water levels observed in paired wells (fig. 11) were 14 to 16 ft higher in the shallow observation wells near the reservoir than in the deep wells. Some shallow wells farther from the reservoir had water levels up to 20 ft higher than nearby deep wells. Most deep wells are screened about 20 to 30 ft deeper than shallow wells. Thus, a large downward vertical gradient exists between the shallow and deep saturated zones underlying the area surrounding the reservoir. The toe of the embankment is a low-permeability transition zone that constrains horizontal flow for both saturated zones.

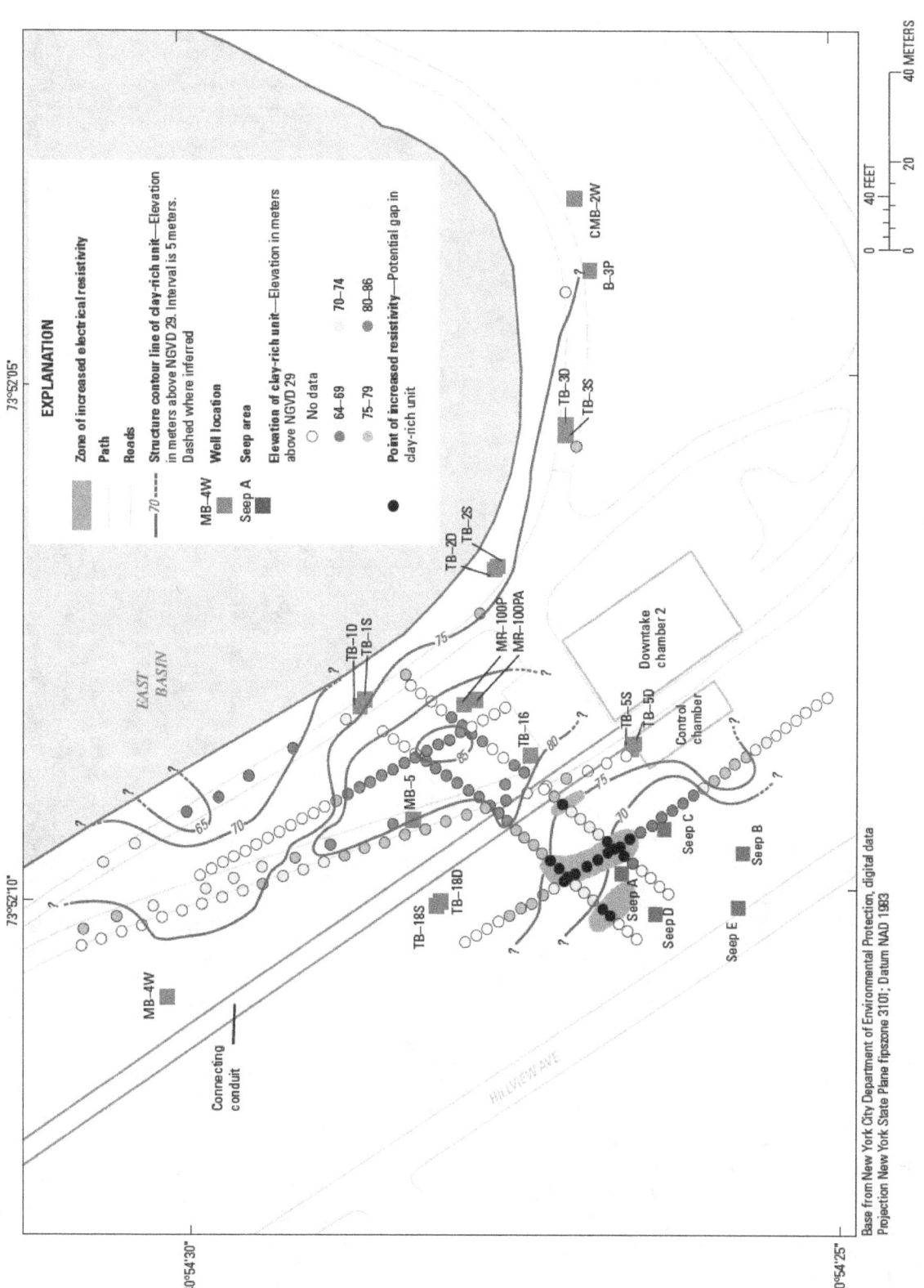

Figure 10. Interpreted elevation of a subsurface clay-rich unit delineated from two-dimensional resistivity and shear wave seismic surveys at the Hillview Reservoir, City of Yonkers, Westchester County, New York. (NGVD 29, National Geodetic Vertical Datum of 1929.)

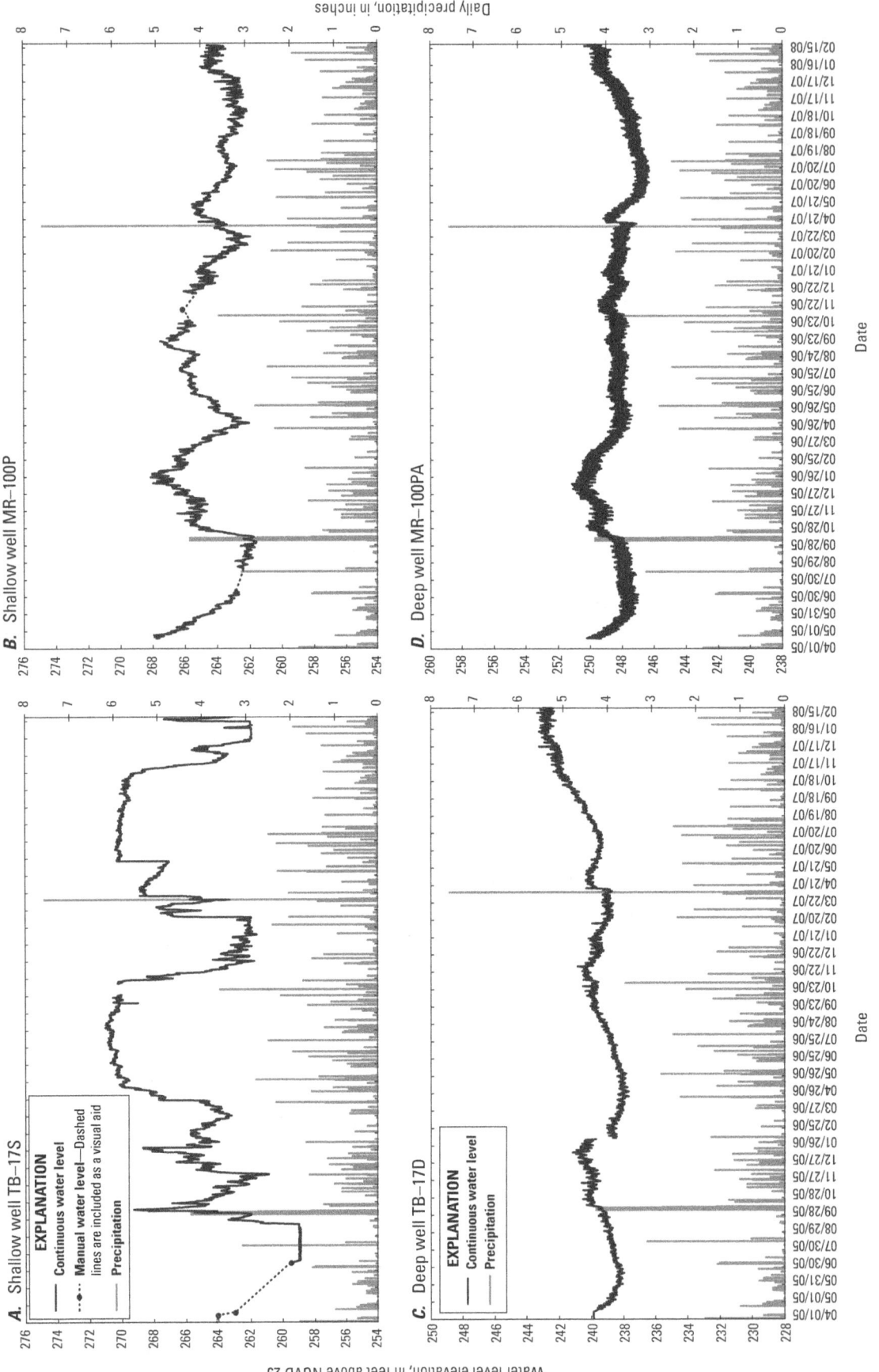

Figure 11. Elevation of water levels in wells A, TB–17S, B, MR–100P, C, TB–17D, and D, MR–100PA in relation to daily precipitation at the Hillview Reservoir, City of Yonkers, Westchester County, New York, between April 1, 2005 and March 1, 2008. Blank where data are missing. Data are from National Oceanic and Atmospheric Administration (2008). NGVD 29, National Geodetic Vertical Datum of 1929.

In general, shallow wells respond fairly quickly to precipitation variations. Exceptions to this observation occurred in wells B–3P, CMB–2W, TB–2S, and TB–17S (fig. 11; appendix 1). Three of these wells, B–3P, CMB–2W, and TB–2S, had among the lowest hydraulic conductivity values of any shallow wells tested (fig. 5; table 3). Since 2003, annual precipitation amounts have exceeded the 138-year average (fig. 4). Groundwater elevations in the shallow water-bearing unit have increased at the Hillview Reservoir accordingly. During the study period from April 2005 through February 2008, water-level elevations in the shallow water-bearing unit appear to have been above average. A water-level contour map of the shallow water-bearing unit for September 2007 indicates that water levels ranged from 286 ft along the reservoir to 192 ft at the toe (fig. 12). Two groundwater mounds, one on the westernmost part and the other near the easternmost part of the study area, are indicated in the contour map. Groundwater appears to be mounding in the vicinities of wells B–3P, CMB–2W, and TB–4S, and wells TB–17S, TB–2S, and MR–100P. These groundwater mounds correlate with areas of low hydraulic conductivity (fig. 5). Water-level elevations appear to be at or above land-surface elevations in the vicinity of the known seeps in the westernmost part of the study area (fig. 12). The localized groundwater mounds indicate that the shallow water-bearing unit is underlain or surrounded by low-permeability materials in these areas. The two surface geophysical surveys delineated a dense, electrically conductive zone underlying the westernmost groundwater mound. This zone is interpreted as being a clay-rich unit (fig. 10) and correlates with the location of an area of low hydraulic conductivity (fig. 5).

The westernmost groundwater mound is larger than the easternmost mound and appears to supply water to seeps A and B. These large groundwater mounds coupled with the steep slopes found at the Hillview Reservoir appear to be the main controlling source of water to the seeps in this part of the reservoir. The lack of a groundwater mound during drier months (lower precipitation) in the vicinity of TB–17S (fig. 2) indicates that the contribution of water from the connecting conduit is limited or has other controlling factors —hydrogeology, precipitation, and reservoir infrastructure. (fig. 11). A leaking connecting conduit in this area might create a consistently above-average water elevation independent of variations in precipitation throughout the study period. The correlation of the emergence of several seeps at the site with the period of above-average precipitation during the past 10 years strongly suggests a connection with the high precipitation and local geologic control. Further study of this area could verify this hypothesis. In general, groundwater tends to flow from the reservoir and the two groundwater mounds outward toward the surrounding glacial overburden. There is also a downward component of groundwater flow into the underlying deep saturated zone at the embankment.

Deep Saturated Zone

Water levels in the deep saturated zone appear to be fluctuating at regular daily intervals as a result of the artificial basin tide. The largest tidal range was observed in wells screened in the deep water-bearing unit closest to the reservoir. Hydrographs of observation wells TB–2D and TB–17D, screened in the deep water-bearing unit, indicate that water levels near the reservoir were strongly affected by the basin, and those about 150 ft from the reservoir were less influenced (fig. 3). Water-level elevations in the deep water-bearing unit were consistently lower than those in the overlying shallow water-bearing unit. Hydrographs of wells screened in the deep water-bearing unit had small seasonal water-level ranges (excluding short term peaks) in the westernmost part of the study area (fig. 11). For example, seasonal water-level ranges for observation wells TB–13 and TB–15 were less than 2 ft (appendix 1). In contrast, water-level fluctuations in the easternmost part of the study area at well TB-12 were greater than 10 ft (appendix 1), which indicates that the materials underlying the easternmost part of the study area may be more permeable than those underlying the westernmost part. In general, the farther away a well is from the reservoir, the greater the range in water levels over the period of study. A water-level contour map of the deep water-bearing unit for September 2007 indicates water levels ranged from 192 to 269 ft (fig. 13). The area along the toe of the embankment is a low-permeability transition zone that constrains flow for both saturated zones. The rapid thinning of both the shallow and deep saturated zones along the toe of the embankment with a shallow depth to bedrock below, forces the groundwater from both zones to flow through this area. Recharge from the reservoir appears to be a major contributor to the deep water-bearing unit with lesser influence from precipitation-induced recharge or recharge from the overlying shallow water-bearing unit. In general, groundwater tends to flow from the reservoir outward toward the surrounding glacial drift.

Temperature as a Tracer

One of the earliest applications of using heat as a tracer in groundwater studies was by Slichter in 1905. Winslow (1962) measured a thermal plume of warm water from a nearby river that was induced to flow toward a pumping well field. Conductive transport of heat occurs only in moving groundwater (Freeze and Cherry, 1979). The seasonal variation in heat transported along with flowing groundwater can be measured and analyzed. The transfer of heat by the circulation of recharged or discharged groundwater is a common phenomenon (Anderson, 2005). In general, the amplitude of temperature fluctuations decreases with depth (Sillman and Booth, 1993). Below about 1.5 m (5 ft) groundwater temperatures are not substantially affected by diurnal temperature fluctuations at land surface.

Groundwater temperature curves were used as an indicator of groundwater-flow velocities at the site along the

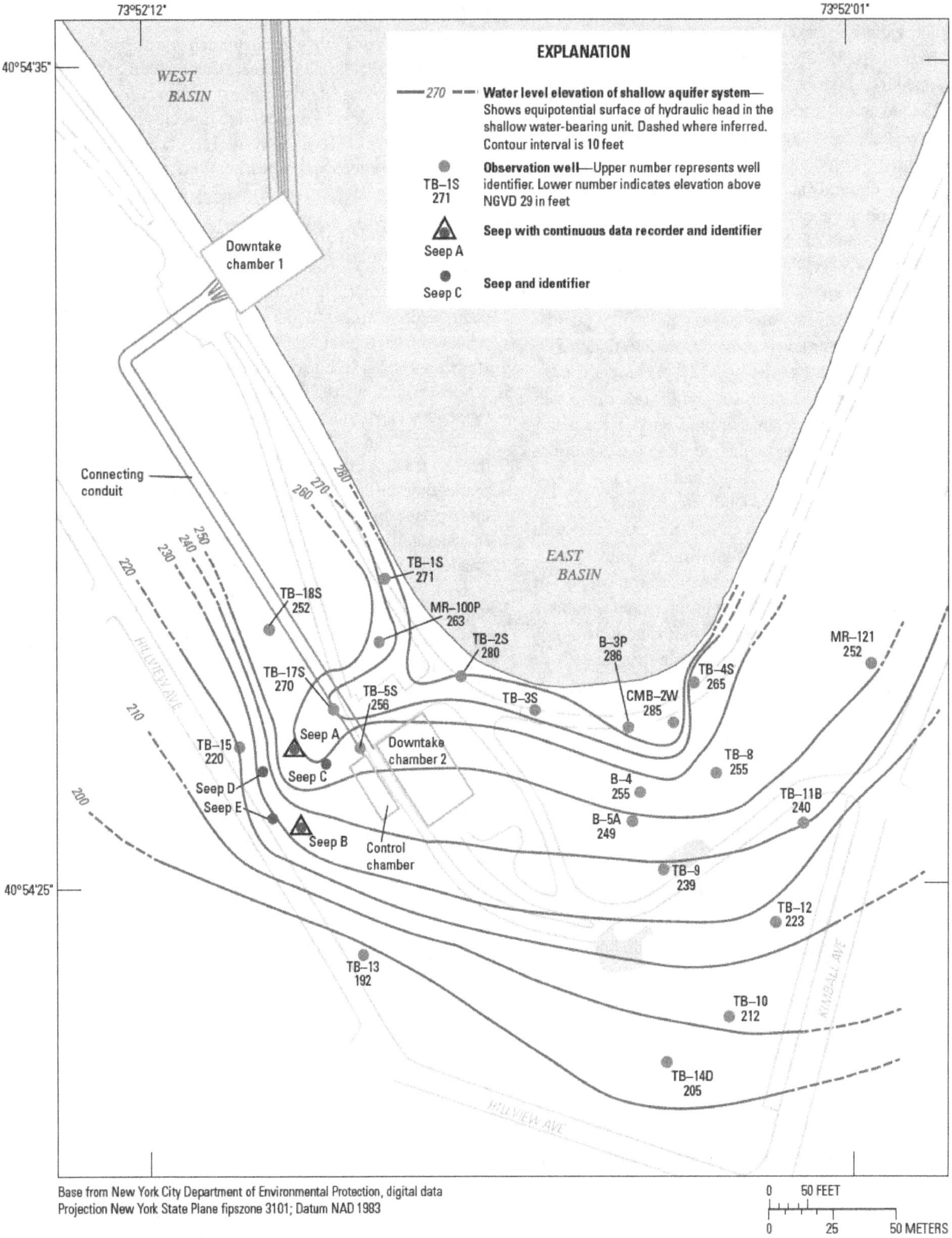

Figure 12. Water-level elevation in the shallow water-bearing unit at the southern embankment of the Hillview Reservoir, City of Yonkers, Westchester County, New York, in September 2007. NGVD 29, National Geodetic Vertical Datum of 1929.

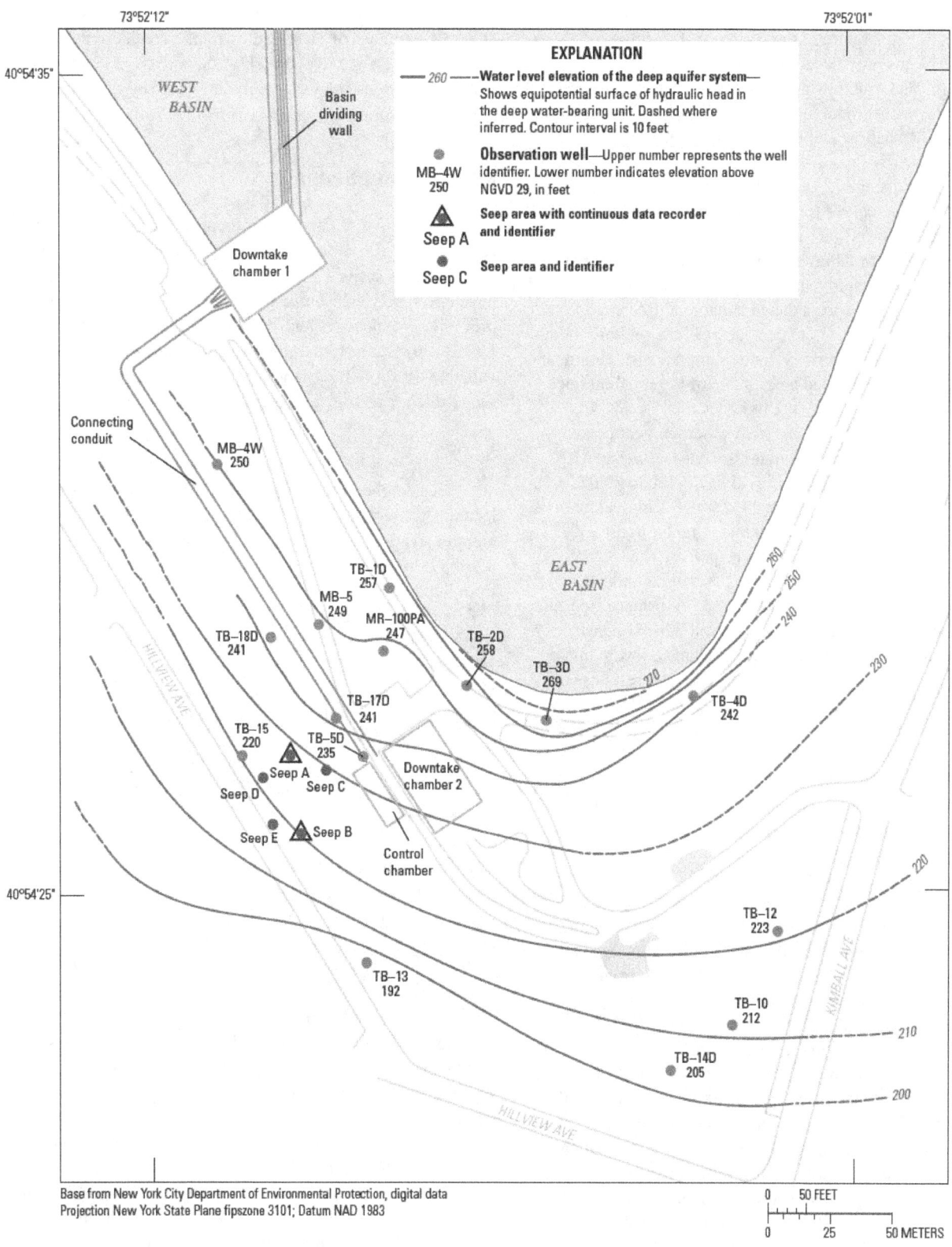

Figure 13. Water-level elevation in the deep water-bearing unit at the southern embankment of the Hillview Reservoir, City of Yonkers, Westchester County, New York, in September 2007. NGVD 29, National Geodetic Vertical Datum of 1929.

150-ft perimeter of the reservoir's East Basin. The seasonal variation in temperature of the surface water within the reservoir basins provides a source of heat for use as a tracer of recharge and groundwater flow at the site. Groundwater contour maps of the site indicate water from the East Basin recharges the shallow and deep saturated zones. Using the temperature curves recorded from the East Basin's digital pressure transducer as a guide, the time delays in the high- and low-temperature curves for wells instrumented with digital pressure transducers were evaluated for wells within 150 ft of the reservoir. Based on hydraulic-head gradients in the shallow and deep saturated zones, groundwater flows outward from the East Basin across the embankment. Recharge from precipitation also appears to be a factor in the shallow water-bearing unit. Limiting this evaluation to the 150-ft perimeter of the East Basin provided an estimate of the apparent linear groundwater-flow velocity in this zone.

The East Basin reached a high water temperature of 18.5 degrees Celsius (°C) in mid-August 2006 and a low temperature of 1.0°C in mid- to late February 2007. Groundwater in well TB–1S reached a high temperature in mid- to late February 2007 and a low in mid-August 2007, a 6-month delay in the groundwater response, which equated to an estimated linear velocity of 4.2 feet per month (ft/mo) (0.14 foot per day (ft/d)) based on the 25-ft distance from the East Basin (fig. 14). Groundwater in well TB–1D reached a high temperature in mid-January 2007 and a low in mid-July 2007, a 5-month delay in the groundwater response or an estimated linear velocity of 5 ft/mo (0.17 ft/d). Similar velocities were estimated for wells TB–2S, TB–2D, TB–3D, TB–4S, CMB–2W, MR–100P, and MR–100PA (table 3). Within a given saturated zone, the temperature curve delays increased as the distance between wells and the East Basin increased. In addition, differences in temperature curve delays were observed between wells screened in the shallow saturated zone and those in the deep saturated zone (fig. 14). In general, the temperature amplitude decreased and the delay increased in the deep water-bearing unit compared with the amplitude and delay in the shallow water-bearing unit. Although wells TB–1S and TB–2D are about the same distance from the East Basin, their temperature curves are different (fig. 14). The temperature data indicate that the estimated linear groundwater velocities in the deep saturated zone at TB–2D are slightly higher than those in the shallow saturated zone at TB–1S. A lack of instrumented wells in this area limited the application of this technique. Based on the temperature data, the estimated groundwater-flow velocities ranged from about 4 to 9 ft/mo (table 3). One exception was for well Z–PA, which had a delay of only 2 months for an apparent velocity of 13 ft/mo. Wells beyond 150 ft from the reservoir had delays that were longer than 18 months and, thus, estimated groundwater-flow velocities could not be accurately determined. However, the wells within the southern part of the study area appear to be relatively similar in apparent linear velocities. There were some indications of an increase in estimated linear velocities in wells farther northwest of MR–100P and in the well pairs

at TB–1 and TB–2. This finding seems to correlate with the slug test estimates of increased hydraulic conductivity in wells northwest of MR–100P, TB–1S, and TB–2S and the location of a dense clay-rich unit delineated by the surface geophysical surveys.

Embankment Seepage

At least five separate seeps located downslope from the reservoir have been documented at the Hillview Reservoir, and are defined as seeps A, B, C, D, and E. Based on the data that have been analyzed, source water to the seeps appears to be primarily groundwater and, to a lesser extent, water from the East Basin. A long-term monitoring network would provide valuable data for understanding hydrologic controls on the seeps at the reservoir. The seeps are described in detail below.

Seep A

Seep A, one of the longest running seeps at the site, is located 85 ft downslope from well TB–17S (fig. 2). The seep has an elevation of 255.3 ft. Manual discharge measurements of seep A by NYCDEP consultants were made sporadically from September through November 2001 and ranged from 0 to 8.3 gallons per minute (gal/min). The seep appears to have continued to flow sporadically through February 2008. In May 2005, the USGS installed a Parshall flume at this seep with a digital pressure transducer to record stage at 1-hour intervals. Stage was recorded digitally using a pressure transducer at hourly intervals in both the flume and weir to calculate discharge. A rating equation was used to calculate discharge from the stage measurements collected at the flume.

Manual measurements were collected to calibrate the digital data. In May 2005, discharge was measured to be about 8 gal/min; thereafter, discharge rapidly declined (fig. 15). Precipitation from May through October 2005 was below average. Discharge at seep A dropped to less than 1 gal/min until November 2005. A sudden increase to about 4.8 gal/min of discharge was measured at seep A in November 2005 through April 2006. This increase in discharge corresponds to the third wettest autumn (2005) and the wettest October ever recorded in Central Park. However, another rapid decrease in discharge occurred during May 2006 that appeared to coincide with the installation of a new french drain system behind the retaining wall downslope from this seep. It appears the new drain system, which does not have a discharge monitor, may have altered the source of water to this seep. Increases in precipitation in 2006 and 2007 no longer caused increases in discharge at this seep. In May 2006, a weir was installed (to replace the flume) with a digital pressure transducer at seep A to provide more accurate low-discharge measurements. A rating equation was used to calculate discharge from the stage measurements collected at the weir. From May 2006 through February 2008, discharge at seep A did not exceed 2 gal/min and averaged about 1 gal/min or less (fig. 15).

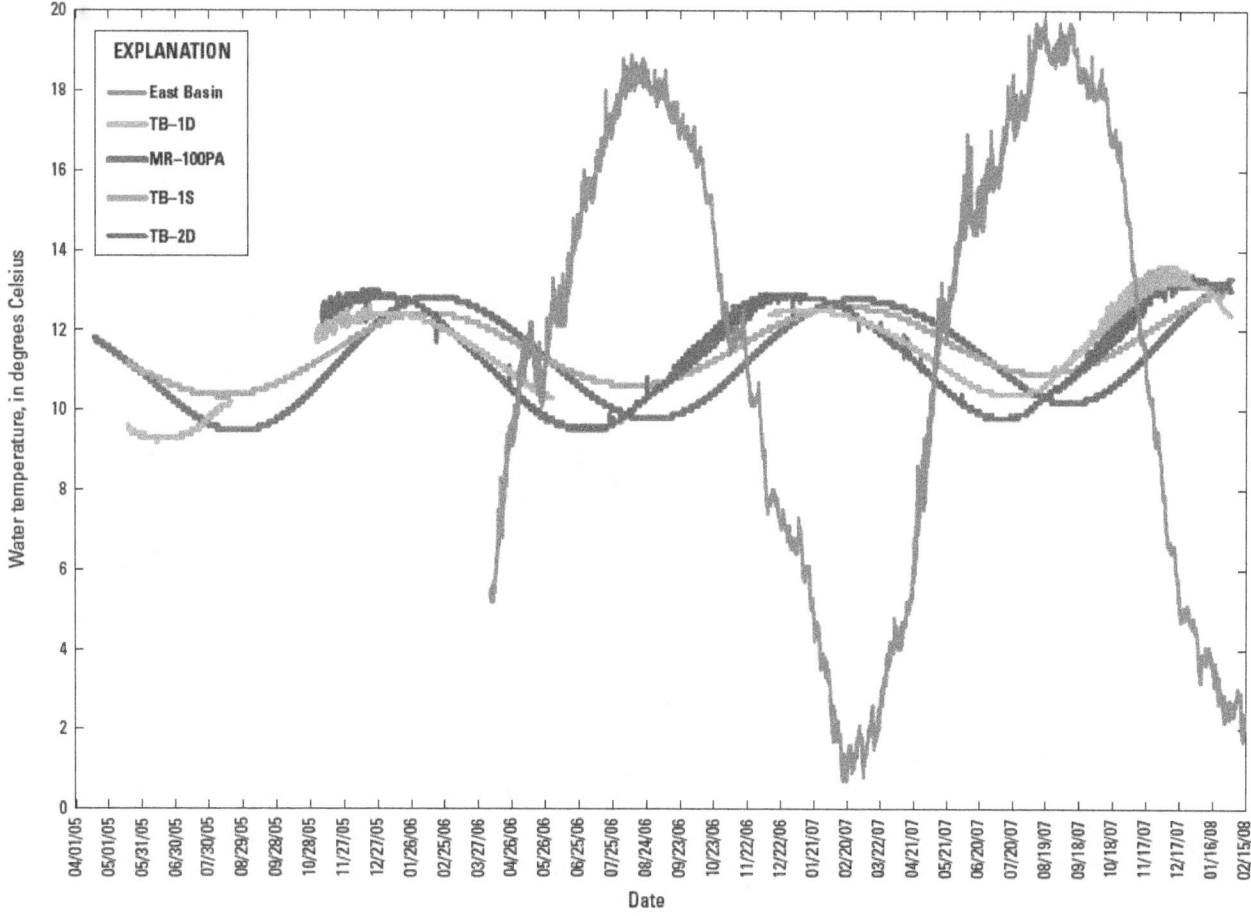

Figure 14. Water temperature in the East Basin and wells TB–1S, TB–1D, TB–2D, and MR–100PA at the Hillview Reservoir, City of Yonkers, Westchester County, New York, between April 1, 2005 and March 1, 2008. Blank where data are missing.

Seep B

Seep B is located 100 ft downslope from the control chamber and about 70 ft upslope from the retaining wall along Hillview Avenue. (fig. 2). During a field visit by the USGS in August 2004, a small stream (seep B) was observed flowing from the steep embankment. On subsequent field visits to the embankment in autumn 2004, the discharge had stopped flowing from seep B. The discharge was not quantified at that time but appeared to resemble discharge amounts observed during visits in 2007. During a field visit in September 2007, when the USGS installed a Parshall flume at this seep, discharge from seep B was observed to be flowing again. This seep appears to flow only after periods of sustained above-average precipitation and high water-level elevations in the shallow saturated zone. The flume was instrumented with a digital pressure transducer to record stage at hourly intervals. The seep was observed to produce large quantities of discharge and formed a small stream during a field visit in August 2004. The seep appears to have stopped flowing near the end of fall 2004; it began flowing again in August 2007.

Manual discharge measurements were as high as 21.4 gal/min. A rating equation was used to calculate discharge. Spring and summer 2007 were the fourth wettest seasons ever recorded at Central Park. April 2007 was the second wettest month on record at Central Park. Discharge from this seep appeared to erode sediment at the toe. Sediment grab samples were obtained near the source of this seep. Discharge from seep B steadily declined through February 2008. From November 2007 to the end of this study on February 1, 2008, discharge averaged 6 gal/min (fig. 15). However, seasonal increases in the discharge of this seep have been observed since February 2008.

Based on observations of seep B, discharge may increase during periods of above-average precipitation and high water-level elevations in the shallow water-bearing unit. Because of the discontinuous nature of this seep, erosion of fine materials from within the embankment may occur during these periods of increased discharge. Results of analysis of sediment grab samples from seep B indicate that the suspended sediment concentration ranged from 1 to 9 mg/L. The suspended sediment size ranged from 31 to 89 percent less than the

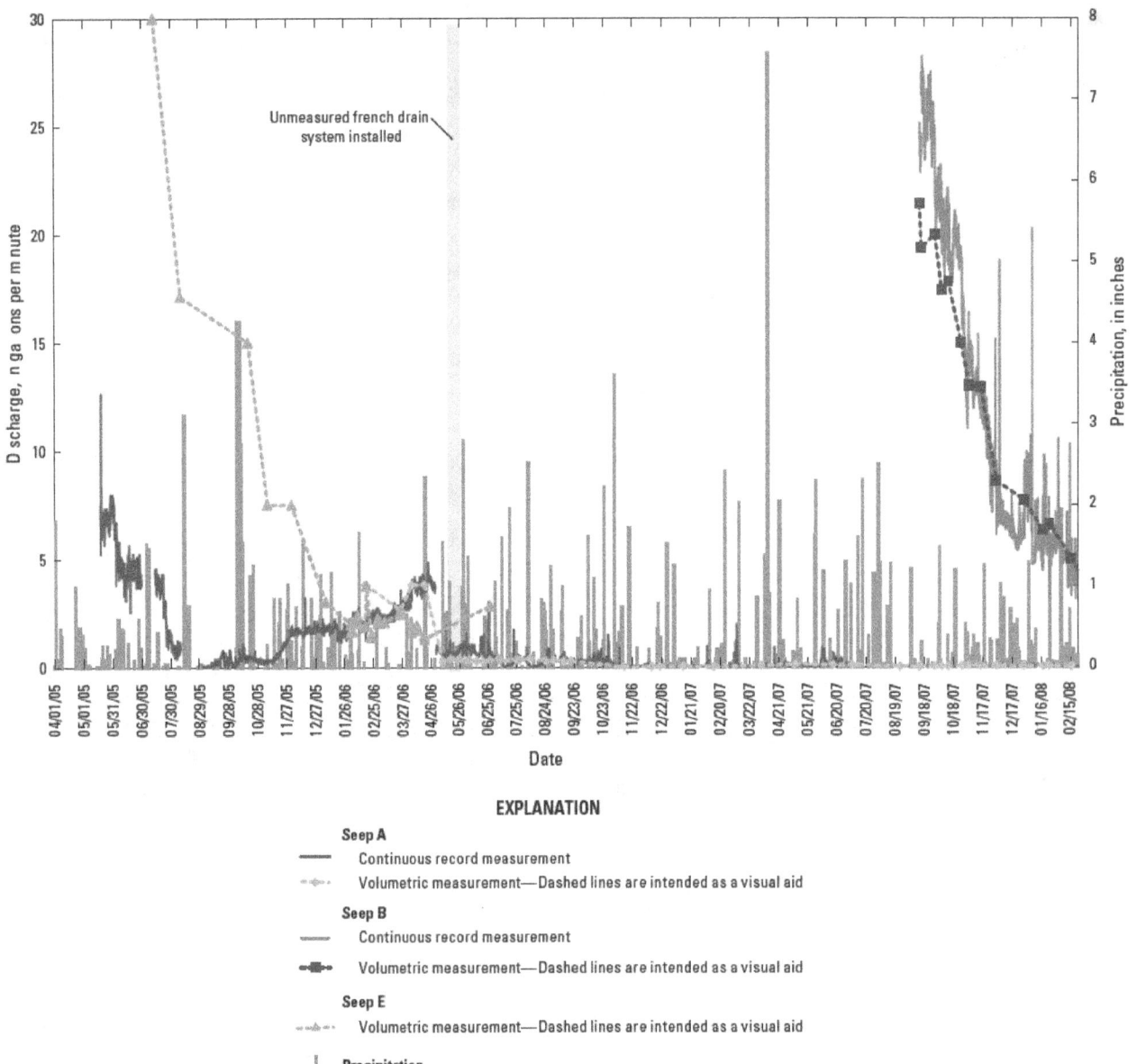

Figure 15. Discharge rates of seeps measured on the southern embankment at the Hillview Reservoir, City of Yonkers, Westchester County, New York, and daily precipitation between April 1, 2005, and March 1, 2008, recorded at Central Park, New York County, New York. Blank where data are missing.

0.063-millimeter (mm)-sieve size (table 4). This finding indicates that most of the sediment load is a fine-grained silt or clay.

Hydrographs indicate an increase in groundwater elevations at two wells completed in the shallow water-bearing unit, TB–17S and TB–2S, near seep B (fig. 11; appendix 1). Increased precipitation in 2007 produced the fourth wettest spring and summer. Increases in water levels in the shallow water-bearing unit were recorded during this period at the two wells. At well TB–2S, water levels were at a historic high, and those at TB–17S, nearly so. A relatively dry autumn

(2007) appears to have lowered groundwater levels in many shallow wells at the southern embankment of the study area. The steady decline in discharge measured at seep B appears to correlate closely with water-level elevations at well TB–17S (fig. 16).

It is unclear why there was no discharge flowing at the seep B site during similar high water levels in TB–17S in 2006. One possible explanation may be that a low permeable zone is present in the vicinity of TB–17S and TB–2S, indicating an impediment to drainage in this area. This low permeable zone was delineated by the surface geophysical

Table 4.   Suspended sediment in surface-water samples from seep B at the Hillview Reservoir, City of Yonkers, Westchester County, New York.

[U.S. Geological Survey (USGS) groundwater spring designated as seep B with USGS identification number 405425073520901. <, less than; mm, millimeter; mg/L, milligrams per liter]

Sample date and time	Sample medium	Suspended sediment, <0.063-mm-sieve diameter, in percent	Suspended sediment concentration, in mg/L	Sampling method
9/6/2007 9:44	Surface water	31	8.0	Grab
9/6/2007 11:07	Surface water	54	9.0	Grab
3/11/2008 13:30	Surface water	89	1.0	Grab
3/11/2008 13:35	Surface water	88	1.0	Grab

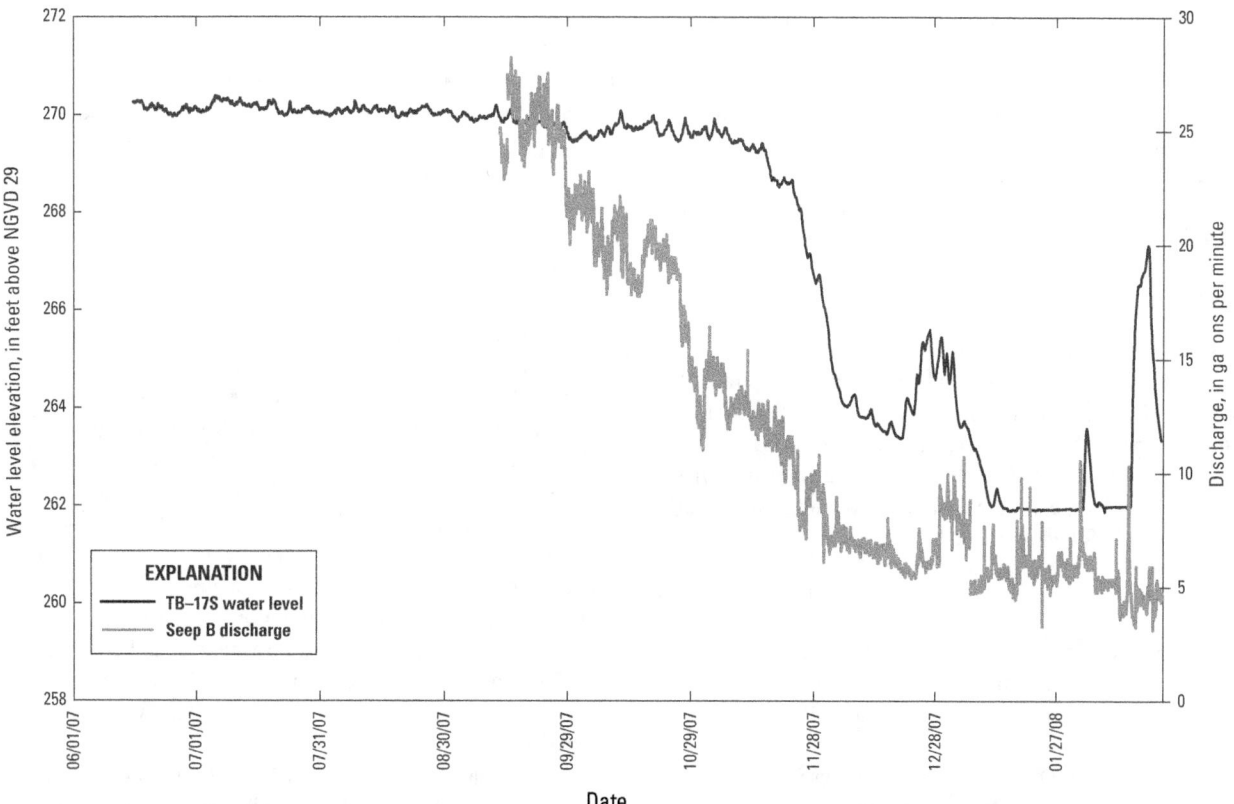

Figure 16.   Groundwater levels in well TB–17S and discharge rates at seep B at the Hillview Reservoir, City of Yonkers, Westchester County, New York, between June 1, 2007 and March 1, 2008. (NGVD 29, National Geodetic Vertical Datum of 1929.) Blank where data are missing.

surveys and hydraulic testing. The low permeable zone correlates with low estimated linear groundwater velocities, calculated using groundwater temperature curves, and the location of a large groundwater mound. The low permeable zone appears to function much like a bowl that fills partially from basin recharge and mostly during periods of above-average precipitation. To further explore the effect that the low permeable zone has on drainage in this area, water-level fluctuations in well TB–17S located in the low permeable zone were compared with water-level fluctuations in well TB–18S located outside the low permeable zone. Evaluation of water levels in TB–18S indicates rapid increases in groundwater elevations after prolonged above-average precipitation (appendix 1). Whereas water levels in TB–17S increase, the elevated water levels in TB–18S appear to decline, and the shallow saturated zone in that area returns to a lower level. The water level at TB–17S returns to a background level at a slow rate (fig. 11). This appears to correlate with geophysical delineation of a bowl-like ridge of dense, electrically conductive material (low-permeable clay) that underlies the westernmost part of the study area. Further study would be required to better understand this mechanism.

Seep C

Seep C, located 30 ft downslope from the control chamber, appears to have started flowing at the same time as seep B (fig. 2). Seep C, at an elevation of 259.8 ft, likely flowed for only a few weeks. Seep C was observed during a field visit in September 2007. Seep C does not have a specific point source and could be better characterized as a saturated soil area. Discharge was difficult to quantify and never exceeded 1 gal/min. Discharge at seep C appears to have stopped in November 2007.

Seep D

Seep D is located 50 ft downslope from seep A, was observed throughout this study and during field visits in 2003. Seep D is characterized by a small area of constantly saturated soil near the base of the embankment about 50 ft downslope from seep A. No discharge has been observed emanating from this seep. Seep D appears to have been a wet area since field visits by the USGS began in 2003.

Seep E

Seep E, encompassing the embankment toe or edge along Hillview Avenue, is the largest seep of a group of seeps at the base of a stone retaining wall that runs along a part of Hillview Avenue at the toe of the embankment downslope of seeps A, B, C, and D. Although most of these seeps can be characterized as small wet areas or minor discharges, one opening in the retaining wall produced large discharges. This seep is expressed as wet areas at the base of a stone embankment retaining wall. One particular area along the retaining wall has an opening from which discharge was

measured manually and ranged from less than 1 to 30 gal/min (fig. 15). During July 2005, a 30-gal/min discharge was measured manually. Subsequent field visits to this seep indicated a rapid decrease in discharge during late spring and summer 2005. In February 2006, the discharge dropped to less than 2 gal/min, and by July 2006, the discharge was below measureable levels. After the installation of the french drainage system behind the retaining wall, no discharge was observed from this seep. However, many wet areas continued to be observed along the retaining wall, indicating a potential buildup of groundwater in the clay-rich sediments there.

Summary

The earthen embankment at the Hillview Reservoir comprises low-permeability clays and glacial till that were excavated from the site; the embankment rests on a thin veneer of glacial-drift deposits that overlie crystalline bedrock. The study area is underlain by two permeable water-bearing units—the shallow and deep saturated zones. In the past, several seeps located downslope from the reservoir have flowed from the steepest slope at the southern end of the earthen embankment including one seep (seep A) that has been flowing continuously since about 1999. In 2001, the New York City Department of Environmental Protection (NYCDEP) installed 25 wells at the southern end of the reservoir (for a total of 57 wells around the reservoir) in an effort to locate the source of the continuous flowing seep. The U.S. Geological Survey (USGS) was able to use 46 of the 57 wells to monitor water levels, temperature, and water quality.

Water-Quality Samples.—Water samples were collected from 12 groundwater observation wells, the East Basin of the reservoir, and two seeps for chemical analysis. Samples were analyzed for nutrients, major ions, metals, pesticides, pesticide degradates, and volatile organic compounds. Results of the water-quality analysis of groundwater samples collected in this study identified arsenic concentrations greater than 1 microgram per liter (μg/L) at 8 of the 15 sites sampled. Well B–3P had the highest concentration of arsenic at 21.8 μg/L. The sample collected from B–3P also had the highest concentrations of metals and chloride of the 15 sites that were sampled. Toluene was detected at concentrations greater than the detection limit at 5 of the 15 sites sampled.

Trace amounts of bromodichloromethane, a trihalomethane, were detected at downtake chamber 1, well TB–17S, and seeps A and E. Using bromodichloromethane as a tracer, there appears to be a connection between the downtake chamber, well TB–17S, and seeps A and E. Trichloromethane was detected at 13 of the 15 sites sampled. With the exception of the TB–17 well pair, the deep wells had the highest trichloromethane concentrations. No trichloromethane was detected at shallow wells TB–1S, B–3P, and TB–18S. Concentrations of trichloromethane in samples collected at well TB–15, downtake chamber 1 (East Basin), and seeps A and E were in the middle of the concentration range of the deep and shallow well samples, which suggest a connection between the downtake chamber and the seeps. Fluoride detected in samples collected from both seeps indicate a connection with water from the reservoir.

Reservoir Tidal Effect.—Evaluation of the hydrograph data indicates that water levels in the wells closest to the East Basin are strongly affected by the artificial tides created by fluctuations in the elevation of the water level in the basin as a result of water use. This cyclic demand produces an artificial "tidal" load on the surrounding embankment materials and local groundwater-flow system. The effect of the basin tides on the water levels in the wells appears to be related to the distance of the wells from the basin and the depth of the wells. In general, the water-level elevation in the basin ranges from 3 to 4 feet (ft) several times a day; the mean water elevation in the basin is 293 ft.

Precipitation and Groundwater Levels.—In 2003, 2004, 2005, 2006, and 2007, the annual precipitation exceeded the 138-year average by 30, 16, 24, 33, and 37 percent, respectively. Between January 2004 and March 2008, 7 months had recorded monthly precipitation totals that were among the top 10 wettest on record for specified months. Five of the top 10 wettest seasons on record also occurred during 2005–07.

All the wells showed a downward trend in water-level elevations during spring 2006. This decline corresponds to a relatively dry period that occurred that spring, which included the driest March in the Central Park, New York, precipitation record. Water levels appeared to rebound in response to increased precipitation in the months that followed the dry period.

Hydraulic Conductivity.—Water-level data from 25 wells were used to estimate hydraulic conductivity in the study area. The results indicate a range of estimated hydraulic conductivities from 0.0063 to 1.2 feet per day (ft/d) and averaged 0.17 ft/d. An area of lower than average hydraulic conductivity was delineated between the TB–17 and TB–3 well pairs and south toward Hillview Avenue along the southwestern part of the earthen embankment. There appears to be an abrupt transition from low permeability units to higher permeability units in the northwestern part of the study area. Two isolated areas of higher than average permeable deposits at wells TB–3D, B–4, and TB–4S also were indicated by the results of slug tests.

Surface Geophysics.—Results of the two-dimensional (2D) resistivity surveys indicate a mound of electrically conductive material beneath the terrace area (top of dam) surrounding the reservoir (crest) with a distinct increase in elevation closer to the crest. The conductive zone appears to decrease in elevation toward the reservoir beneath the terrace and downslope toward Hillview Avenue.

Results of the 2D shear wave velocity surveys indicate a similar structure with a zone of high velocity and increasing in elevation toward the crest and decreasing in elevation toward the reservoir and the northern part of the study area.

The bedrock was imaged along a northwest-southeast trending line (G–G′) west of the reservoir; the bedrock ranged from 35 meters (m) (114.8 feet (ft)) below land surface near well B–3P to 36 m (118.1 ft) below land surface at well TB–1D. The high-velocity zone averaged almost 200 meters per second (656.2 feet per second) greater than the overlying material. This high-velocity zone ranges from only 6 m (19.7 ft) below land surface (BLS) near MB–5 to almost 28 m (91.9 ft) BLS north of the TB–1 well pair. When compared, the low-resistivity zone identified with the 2D resistivity survey and the high shear wave velocity zone appear to correlate within a few meters, indicating that a zone of high conductivity and high shear wave velocity underlies the westernmost part of the study area. This zone may be a well-consolidated unit with higher clay content than the surrounding material.

Hydrologic Conditions.—Groundwater elevations were measured manually at 46 observation wells during the study and digitally monitored at selected wells at hourly intervals. Water levels in the shallow zone (upper 45 ft of embankment) are the highest in the study area and appear to be affected by precipitation.

Water levels in shallow observation wells near the reservoir averaged 14 to 16 ft higher in elevation than those in the deep wells at well pairs. Some shallow wells farther from the reservoir had water-level elevations up to 20 ft higher than those in nearby deep wells. A large downward vertical gradient exists between the shallow and deep saturated zones in the area surrounding the reservoir. The toe (edge) of the embankment represents a constrained flow transition zone for both saturated zones.

Two groundwater mounds, one in the westernmost part and the other near the easternmost part of the study area, correlate with areas of low hydraulic conductivity. Groundwater elevations are at or above land surface in the vicinity of the known seeps in the westernmost part of the study area. A dense, electrically conductive zone (clay) that underlies the westernmost groundwater mound correlates with an area of low hydraulic conductivity. The westernmost groundwater mound is larger than the easternmost mound and appears to supply water to seeps A and B. These large groundwater mounds coupled with the steep slopes found at the southern end of the Hillview Reservoir appear to be the source of the seeps in this part of the reservoir. The lack of a groundwater mound near some wells during drier months indicates the contribution of water from the connecting conduit is either limited or has other controlling factors— hydrogeology, precipitation, and reservoir infrastructure.

Temperature as a Tracer.—Delays in groundwater-temperature responses between the East Basin and the groundwater observation wells were used as an indicator of groundwater-flow velocity. Based on hydraulic-head gradients in both the shallow and deep saturated zones, flow appears to move from the reservoir outward across the embankment. A 6-month delay in groundwater flow (which equates to an estimated linear velocity of 4.2 feet per month (ft/mo)

(0.14 foot per day (ft/d)) based on the 25-ft distance from the East Basin) was calculated between the East Basin and well TB–1S, the shallow well in the TB-1 well pair. Similarly, a 5-month delay or a linear velocity of 5 ft/mo (0.17 ft/d) was calculated between the East Basin and well TB–1D, the deep well in the TB-1 well pair. Similar velocities were found at TB–2S, TB–2D, TB–3D, TB–4S, CMB–2W, MR–100P, and MR–100PA. As the distances between the East Basin and the wells increased, temperature curve delays increased.

Seeps.—At least five separate seeps have been documented at the Hillview Reservoir. The five seeps are defined as seeps A, B, C, D, and E.

Seep A is one of the longest continuous running seeps at the site and is located 85 ft downslope from well TB–17S at an elevation of 255.3 ft. Discharge from seep A was measured from September through November 2001. The seep appears to have continued to flow sporadically through February 2008. Precipitation appeared to be a substantial contributing factor to flow from seep A until a french drain system was installed downslope from the seep in May 2006. It appears the new drain system may have affected the flow from seep A; discharge at the seep did not exceed 2 gallons per minute (gal/min) after the drain system was installed.

Seep B is located 100 ft downslope from the control chamber and appears to flow only after periods of sustained above-average precipitation and high water-level elevations in the shallow saturated zone. Because of the discontinuous nature of discharge from this seep, erosion of fine materials from within the embankment may occur during periods of increased discharge. The steady decline in discharge measured at seep B appears to correlate closely with the water-level elevations in well TB–17S. It is unclear why no discharge was flowing at seep B during similar high water levels in well TB–17S in 2006. The seep stopped flowing near the end of fall 2004 and began to flow again in August 2007. Manual discharge measurements were as high as 21.4 gal/min.

Seep C appears to have started at the same time as seep B and is located 30 ft downslope from the control chamber. Seep C was observed during a field visit in September 2007. Discharge from this seep averaged less than 1 gal/min. The seep appears to have stopped in November 2007. The elevation of seep C is 259.8 ft.

Seep D is characterized by a small area of constantly saturated soil near the base of the embankment about 50 ft downslope from seep A. No discharge has been observed emanating from this seep. Seep D appears to have been a wet area since field visits by the USGS began in 2003.

Seep E is the largest seep of a group of seeps coming from the base of a stone retaining wall that runs along a part of Hillview Avenue at the toe of the embankment downslope from seeps A, B, C, and D. Although most of these seeps can be characterized as small wet areas or minor discharges, one opening in the retaining wall produced large discharges. During July 2005, discharge was manually measured to be 30 gal/min.

The hydrologic factor that appears to be controlling the seeps is above-average precipitation, which has a substantial effect on water levels in the shallow saturated zone. When the precipitation recharges the shallow saturated zone, the groundwater flows primarily through the higher hydraulic conductivity materials away from the reservoir. Based on the data that have been analyzed, source water to the seeps appears to be primarily groundwater and, to a lesser extent, water from the East Basin. A long-term monitoring network would provide valuable data for understanding hydrologic controls on the seeps at the reservoir.

References Cited

Anderson, M.P., 2005, Heat as a ground water tracer: Ground Water, v. 43, no. 6, p. 951–968.

Asselstine, E.S., and Grossman, I.G., 1955, The groundwater resources of Westchester County, New York, part 1 of Records of wells and test holes: New York State Water Power and Control Commission Bulletin GW–35, 79 p.

Baskerville, C.A., 1982, Adoption of the name Hutchinson River Group and its subdivisions in Bronx and Westchester Counties, southeastern New York, *in* Stratigraphic notes, 1980–1982: U.S. Geological Survey Bulletin 1529–H, p. H1–H10.

Baskerville, C.A., 1992, Bedrock and engineering geologic maps of Bronx County and parts of New York and Queens Counties, New York: U.S. Geological Survey Miscellaneous Investigations Series, map I–2003, 2 sheets, scale 1:24,000.

Bouwer, Herman, and Rice, R.C., 1976, A slug test for determining hydraulic conductivity of unconfined aquifers with completely or partially penetrating wells: Water Resources Research, v. 12, no. 3, p. 423–428.

Freeze, R.A., and Cherry, J.A., 1979, Groundwater: Englewood Cliffs, New Jersey, Prentice Hall, Inc., 604 p.

Halford, K.J., and Kuniansky, E.L., 2002, Documentation of spreadsheets for the analysis of aquifer-test and slug-test data: U.S. Geological Survey Open File Report 02–197, 51 p.

Malcolm Pirnie, Inc., and TAMS Consultants, Inc., 2002, Capital project W–10, Hillview reservoir—Cover, phase I, dividing wall stability buttress construction; south embankment monitoring and leakage investigations: U.S. Environmental Protection Agency, 18 p.

National Oceanic and Atmospheric Administration, 2008, Monthly and annual precipitation at Central Park, Manhattan, New York: National Oceanic and Atmospheric Administration, accessed May 15, 2008, at http://www.erh.noaa.gov/okx/climate/records/monthannualpcpn.html.

Park, C.B., Miller, R.D., and Xia, Jianghai, 1999, Multichannel analysis of surface waves (MASW): Geophysics, v. 64, no. 3, p. 800–808.

Sillman, S.E., and Booth, D.F., 1993, Analysis of time series measurements of sediment temperature for identification of gaining versus losing portions of Juday Creek, Indiana: Journal of Hydrology, v. 146, p. 131–148.

Slichter, C.S., 1905, Field measurements of the rate of movement of underground waters: U.S. Geological Survey Water-Supply and Irrigation Paper no. 140, 122 p.

Stumm, Frederick, Chu, Anthony, Como, M.D., and Noll, M.L., 2012, Preliminary analysis of the hydrologic effects of temporary shutdowns of the Rondout-West Branch water tunnel on the groundwater-flow system in Wawarsing, New York: U.S. Geological Survey Scientific Investigations Report 2012–5015, 48 p., at http://pubs.usgs.gov/sir/2012/5015/.

U.S. Consumer Product Safety Commission, 2011, Guidance for outdoor wooden structures: U.S. Consumer Product Safety Commission Publication 270, 2 p.

U.S. Geological Survey, variously dated, National field manual for the collection of water-quality data: U.S. Geological Survey Techniques of Water-Resources Investigations, book 9, chap. A1–A9, variously paged.

Winslow, J.D., 1962, Effect of stream infiltration on ground-water temperatures near Schenectady, New York, article 111 *in* Geological survey research 1962, Short papers in geology, hydrology, and topography, Articles 60–119: U.S. Geological Survey Professional Paper 450–C, p. C125–128.

Zohdy, A.A., Eaton, G.P., and Mabey, D.R., 1974, Application of surface geophysics to groundwater investigations: U.S. Geological Survey Techniques of Water-Resource Investigations 2–D1, 116 p.

Table 2. Concentrations of detected constituents in water samples collected from the Hillview Reservoir, City of Yonkers, Westchester County, New York, in 2006.

[Numbers in parentheses are National Water Information System (NWIS) parameter codes. BLS, below land surface; µS/cm, microsiemens per centimeter; C, degrees Celsius; mg/L, milligrams per liter; DCPA, dimethyl tetrachloroterephthalate; e, estimated; EPTC, S-ethyl dipropylthiocarbamate; µg/L, micrograms per liter; N, nitrogen; P, phosphorus; SiO$_2$, silicon dioxide; µm, micrometer; <, less than; —, no data]

Local well name	Sampling date	Depth to water level, in feet BLS (72019)	pH, water, unfiltered, field, standard units (00400)	Specific conductance, water, unfiltered, µS/cm (00095)	Temperature, water, in °C (00010)	Elevation of land surface, in feet (72000)	Depth of well, in feet BLS (72008)	Dissolved solids dried at 180 °C, water, filtered, in mg/L (70300)	Calcium, water, unfiltered, recoverable, in mg/L (00916)	Magnesium, water, unfiltered, recoverable, in mg/L (00927)	Potassium, water, unfiltered, recoverable, in mg/L (00937)
East Basin	2/8/2006	—	7.3	103	4.1	—	—	55	5.45	1.29	0.603
Seep A	2/6/2006	—	7.2	393	5.4	255	—	206	13.9	1.72	2.64
Seep E	2/8/2006	—	6.6	356	7	220	—	199	18.4	2.82	2.37
TB–17S	2/7/2006	37	7.4	73	—	297	40	398	42.9	15.9	9.86
TB–5S	5/1/2006	48	7.8	1,850	11.7	299	51	11,300	217	16.8	70.8
B–3P	5/1/2006	13	8	4,070	12.2	301	23	3,060	185	120	55.7
TB–12	5/1/2006	21	7.7	2,350	15.5	247	52	1,270	43.1	9.93	4.33
TB–15	5/2/2006	10	7.1	440	13.1	228	33	240	34.4	11.2	1.43
TB–18D	5/1/2006	36	6.8	2,430	13.2	275	60	988	174	62.3	8.46
TB–1S	2/7/2006	30	7.1	1,140	—	301	40	655	127	64.6	5.25
TB–18S	5/2/2006	17	6.8	1,520	10.2	275	27	965	202	41.8	4.62
TB–1D	2/7/2006	44	6.7	95	12.1	300	122	49	6.46	1.58	0.85
TB–5D	2/6/2006	65	7	155	9.9	299	77	790	82.9	10.5	1.82
TB–2D	2/7/2006	43	7.6	103	11.1	300	71	54	5.92	1.31	0.724
TB–17D	2/7/2006	60	7	195	10.2	297	80	869	138	19.2	3.27

Table 2   35

Table 2. Concentrations of detected constituents in water samples collected from the Hillview Reservoir, City of Yonkers, Westchester County, New York, in 2006.—Continued

[Numbers in parentheses are National Water Information System (NWIS) parameter codes. DCPA, dimethyl tetrachloroterephthalate; e, estimated; EPTC, S-ethyl dipropylthiocarbamate; BLS, below land surface datum; mg/L, milligrams per liter; µg/L, micrograms per liter; µS/cm, microsiemens per centimeter at 25 degrees Celsius (C); N, nitrogen; P, phosphorus; SiO$_2$, silicon dioxide; µm, micrometer; <, less than]

Local well name	Sampling date	Sodium, water, unfiltered, recoverable, in mg/L (00929)	Chloride, water, filtered, in mg/L (00940)	Fluoride, water, filtered, in mg/L (00950)	Silica, water, filtered, in mg/L as SiO$_2$ (00955)	Sulfate, water, filtered, in mg/L (00945)	Ammonia, water, filtered, in mg/L as N (00608)	Nitrate plus nitrite, water, filtered, in mg/L as N (00631)	Nitrite, water, filtered, in mg/L as N (00613)	Orthophosphate, water, filtered, in mg/L as P (00671)	Arsenic, water, unfiltered, in µg/L (01002)
East Basin	2/8/2006	8.71	9.93	1	3.6	7.43	<0.04	0.27	<0.008	<0.02	0.37
Seep A	2/6/2006	56	86.5	0.82	4	14.3	<0.04	0.28	<0.008	<0.02	0.54
Seep E	2/8/2006	42	68.9	0.77	4.5	13.2	<0.04	0.23	<0.008	e0.01	0.27
TB–17S	2/7/2006	106	144	1.37	7.6	20.5	<0.04	0.51	<0.008	0.07	3
TB–5S	5/1/2006	3,790	<1.00	0.27	12.5	77.9	<0.04	2.16	<0.008	0.33	3.1
B–3P	5/1/2006	2,710	1,630	0.32	e0.01	34.1	0.08	0.24	0.015	0.35	21.8
TB–12	5/1/2006	452	652	0.31	<0.2	23.7	e0.02	1.91	<0.008	e0.01	1.5
TB–15	5/2/2006	31.8	77.7	0.15	8.3	13.2	0.09	0.25	<0.008	<0.02	0.99
TB–18D	5/1/2006	190	398	0.16	<0.2	31.1	0.21	0.22	<0.008	e0.01	5.8
TB–1S	2/7/2006	34	142	0.18	17.9	15.7	0.22	e0.05	e0.006	e0.01	8.1
TB–18S	5/2/2006	39.1	258	0.14	14.4	50.8	<0.04	5.11	<0.008	e0.01	1.2
TB–1D	2/7/2006	8.51	9.44	0.9	4.4	5.93	<0.04	0.31	<0.008	<0.02	0.29
TB–5D	2/6/2006	208	367	0.31	8.5	19.8	0.05	0.26	e0.006	<0.09	0.49
TB–2D	2/7/2006	7.64	10.4	0.84	3.7	6.72	e0.03	0.3	<0.008	e0.01	0.36
TB–17D	2/7/2006	195	408	0.32	8.9	52.1	0.07	<0.06	<0.008	0.02	1.5

Table 2. Concentrations of detected constituents in water samples collected from the Hillview Reservoir, City of Yonkers, Westchester County, New York, in 2006.—Continued

[Numbers in parentheses are National Water Information System (NWIS) parameter codes. DCPA, dimethyl tetrachloroterephthalate; e, estimated; EPTC, S-ethyl dipropylthiocarbamate; BLS, below land surface datum; mg/L, milligrams per liter; µg/L, micrograms per liter; µS/cm, microsiemens per centimeter at 25 degrees Celsius (C); N, nitrogen; P, phosphorus; SiO$_2$, silicon dioxide; µm, micrometer; <, less than]

Local well name	Sampling date	Barium, water, unfiltered, recoverable, in µg/L (01007)	Cadmium, water, unfiltered, in µg/L (01027)	Chromium, water, unfiltered, recoverable, in µg/L (01034)	Copper, water, unfiltered, recoverable, in µg/L (01042)	Iron, water, unfiltered, recoverable, in µg/L (01045)	Lead, water, unfiltered, recoverable, in µg/L (01051)	Manganese, water, unfiltered, recoverable, in µg/L (01055)	Mercury, water, unfiltered, recoverable, in µg/L (71900)	Selenium, water, unfiltered, in µg/L (01147)	Silver, water, unfiltered, recoverable, in µg/L (01077)
East Basin	2/8/2006	15.4	<0.040	0.15	0.9	39.8	0.24	14.3	<0.010	e0.050	<0.160
Seep A	2/6/2006	26.2	<0.040	0.61	2	251	0.42	6.2	e0.008	0.18	<0.160
Seep E	2/8/2006	20.7	e0.022	0.42	1.2	55.1	0.38	1.5	<0.010	0.2	<0.160
TB-17S	2/7/2006	283	0.271	71.8	71.2	50,100	26.5	1,710	0.633	0.83	0.35
TB-5S	5/1/2006	495	0.233	61.2	30.9	13,900	9.28	248	0.13	1.79	e0.152
B-3P	5/1/2006	658	0.83	132	276	102,000	123	3,370	0.65	1.14	1.38
TB-12	5/1/2006	158	1.53	13	18.1	7,370	7.58	159	0.032	1.02	0.178
TB-15	5/2/2006	24.5	e0.022	3.4	2.4	3,760	0.79	579	0.021	0.12	<0.160
TB-18D	5/1/2006	310	0.275	163	89.5	41,900	18.8	5,840	0.085	0.38	e0.152
TB-1S	2/7/2006	305	0.391	31.8	46	25,100	22.2	4,190	0.047	0.43	e0.141
TB-18S	5/2/2006	134	0.226	16.7	11.4	5,130	2.85	97.1	0.015	0.61	<0.160
TB-1D	2/7/2006	16.1	<0.040	3.7	2.8	1,120	1.08	32.6	<0.010	0.08	<0.160
TB-5D	2/6/2006	83	e0.029	2.9	2.9	1,210	0.47	1,210	<0.010	e0.070	<0.160
TB-2D	2/7/2006	10.7	<0.040	2.2	1.9	328	1.11	16.2	<0.010	e0.060	<0.160
TB-17D	2/7/2006	96.9	0.105	6.8	10.4	3,520	13.7	3,550	0.011	0.13	<0.160

Table 2. Concentrations of detected constituents in water samples collected from the Hillview Reservoir, City of Yonkers, Westchester County, New York, in 2006.—Continued

[Numbers in parentheses are National Water Information System (NWIS) parameter codes. DCPA, dimethyl tetrachloroterephthalate; e, estimated; EPTC, S-ethyl dipropylthiocarbamate; BLS, below land surface datum; mg/L, milligrams per liter; µg/L, micrograms per liter; µS/cm, microsiemens per centimeter at 25 degrees Celsius (C); N, nitrogen; P, phosphorus; SiO$_2$, silicon dioxide; µm, micrometer; <, less than]

Local well name	Sampling date	Zinc, water, unfiltered, recoverable, in µg/L (01092)	1,2,3-Trichloropropane, water, unfiltered, recoverable, in µg/L (77443)	1,2-Dibromo-3-chloropropane, water, unfiltered, recoverable, in µg/L (82625)	1,2-Dibromoethane, water, unfiltered, recoverable, in µg/L (77651)	1,2-Dichloroethane, water, unfiltered, recoverable, in µg/L (32103)	1,2-Dichloropropane, water, unfiltered, recoverable, in µg/L (34541)	1,3-Dichloropropane, water, unfiltered, recoverable, in µg/L (77173)	1,4-Dichlorobenzene, water, unfiltered, recoverable, in µg/L (34571)	1-Naphthol, water, filtered (0.7 µm glass fiber filter), recoverable, in µg/L (49295)	2,6-Diethylaniline, water, filtered (0.7 µm glass fiber filter), recoverable, in µg/L (82660)
East Basin	2/8/2006	<2	<0.18	<0.51	<0.036	<0.13	<0.029	<0.06	<0.034	<0.0882	<0.0060
Seep A	2/6/2006	4	<0.18	<0.51	<0.036	<0.13	<0.029	<0.06	<0.034	<0.0882	<0.0060
Seep E	2/8/2006	2	<0.18	<0.51	<0.036	<0.13	<0.029	<0.06	<0.034	<0.0882	<0.0060
TB–17S	2/7/2006	116	<0.18	<0.51	<0.036	<0.13	<0.029	<0.06	<0.034	<0.0882	<0.0060
TB–5S	5/1/2006	44	<0.18	<0.51	<0.036	<0.13	<0.029	<0.06	<0.034	<0.0882	<0.0060
B–3P	5/1/2006	415	<0.18	<0.51	<0.036	<0.13	<0.029	<0.06	<0.034	<0.0882	<0.0060
TB–12	5/1/2006	40	<0.18	<0.51	<0.036	<0.13	<0.029	<0.06	<0.034	<0.0882	<0.0060
TB–15	5/2/2006	5	<0.18	<0.51	<0.036	<0.13	<0.029	<0.06	<0.034	<0.0882	<0.0060
TB–18D	5/1/2006	226	<0.18	<0.51	<0.036	<0.13	<0.029	<0.06	<0.034	<0.0882	<0.0060
TB–1S	2/7/2006	73	<0.18	<0.51	<0.036	<0.13	<0.029	<0.06	<0.034	<0.0882	<0.0060
TB–18S	5/2/2006	14	<0.18	<0.51	<0.036	<0.13	<0.029	<0.06	<0.034	<0.0882	<0.0060
TB–1D	2/7/2006	4	<0.18	<0.51	<0.036	<0.13	<0.029	<0.06	<0.034	<0.0882	<0.0060
TB–5D	2/6/2006	3	<0.18	<0.51	<0.036	<0.13	<0.029	<0.06	<0.034	<0.0882	<0.0060
TB–2D	2/7/2006	3	<0.18	<0.51	<0.036	<0.13	<0.029	<0.06	<0.034	<0.0882	<0.0060
TB–17D	2/7/2006	22	<0.18	<0.51	<0.036	<0.13	<0.029	<0.06	<0.034	<0.0882	<0.0060

Table 2   37

Table 2. Concentrations of detected constituents in water samples collected from the Hillview Reservoir, City of Yonkers, Westchester County, New York, in 2006.—Continued

[Numbers in parentheses are National Water Information System (NWIS) parameter codes. DCPA, dimethyl tetrachloroterephthalate; e, estimated; EPTC, S-ethyl dipropylthiocarbamate; BLS, below land surface datum; mg/L, milligrams per liter; μg/L, micrograms per liter; μS/cm, microsiemens per centimeter at 25 degrees Celsius (C); N, nitrogen; P, phosphorus; SiO_2, silicon dioxide; μm, micrometer; <, less than]

Local well name	Sampling date	2-Chloro-2',6'-diethyl-acetanilide, water, filtered, recoverable, in μg/L (61618)	2-Chloro-4-isopropyl-amino-6-amino-s-triazine, water, filtered, recoverable, in μg/L (04040)	2-Ethyl-6-methyl-aniline, water, filtered, recoverable, in μg/L (61620)	3,4-Dichloro-aniline, water, filtered, recoverable, in μg/L (61625)	3,5-Dichloro-aniline, water, filtered, recoverable, in μg/L (61627)	3-Chloro-propene, water, unfiltered, recoverable, in μg/L (78109)	4-Chloro-2-methyl-phenol, water, filtered, recoverable, in μg/L (61633)	Acetochlor, water, filtered, recoverable, in μg/L (49260)	Acrylonitrile, water, unfiltered, recoverable, in μg/L (34215)	Alachlor, water, filtered, recoverable, in μg/L (46342)
East Basin	2/8/2006	<0.005	e0.005	<0.004	<0.0045	—	<0.50	<0.0056	<0.006	<0.80	<0.005
Seep A	2/6/2006	<0.005	e0.004	<0.004	<0.0045	—	<0.50	<0.0056	<0.006	<0.80	<0.005
Seep E	2/8/2006	<0.005	e0.004	<0.004	<0.0045	—	<0.50	<0.0056	<0.006	<0.80	<0.005
TB-17S	2/7/2006	<0.005	e0.004	<0.004	<0.0045	<0.012	<0.50	<0.0056	<0.006	<0.80	<0.005
TB-5S	5/1/2006	<0.006	e0.006	<0.010	<0.0045	<0.012	<0.50	<0.0050	<0.006	<0.80	<0.005
B-3P	5/1/2006	<0.006	e0.010	<0.010	<0.0045	<0.012	<0.50	<0.0050	<0.006	<0.80	<0.005
TB-12	5/1/2006	<0.006	e0.007	<0.010	<0.0045	<0.012	<0.50	<0.0050	<0.006	<0.80	<0.005
TB-15	5/2/2006	<0.006	e0.006	<0.010	<0.0045	<0.012	<0.50	<0.0050	<0.006	<0.80	<0.005
TB-18D	5/1/2006	<0.006	e0.006	<0.010	<0.0045	<0.012	<0.50	<0.0050	<0.006	<0.80	<0.005
TB-1S	2/7/2006	<0.005	e0.006	<0.004	<0.0045	—	<0.50	<0.0056	<0.006	<0.80	<0.005
TB-18S	5/2/2006	<0.006	<0.014	<0.010	<0.0045	<0.012	<0.50	<0.0050	<0.006	<0.80	<0.005
TB-1D	2/7/2006	<0.005	e0.005	<0.004	<0.0045	—	<0.50	<0.0056	<0.006	<0.80	<0.005
TB-5D	2/6/2006	<0.005	e0.005	<0.004	<0.0045	—	<0.50	<0.0056	<0.006	<0.80	<0.005
TB-2D	2/7/2006	<0.005	<0.006	<0.004	<0.0045	—	<0.50	<0.0056	<0.006	<0.80	<0.005
TB-17D	2/7/2006	<0.005	<0.006	<0.004	<0.0045	—	<0.50	<0.0056	<0.006	<0.80	<0.005

Table 2. Concentrations of detected constituents in water samples collected from the Hillview Reservoir, City of Yonkers, Westchester County, New York, in 2006.—Continued

[Numbers in parentheses are National Water Information System (NWIS) parameter codes. DCPA, dimethyl tetrachloroterephthalate; e, estimated; EPTC, S-ethyl dipropylthiocarbamate; BLS, below land surface datum; mg/L, milligrams per liter; µg/L, micrograms per liter; µS/cm, microsiemens per centimeter at 25 degrees Celsius (C); N, nitrogen; P, phosphorus; SiO_2, silicon dioxide; µm, micrometer; <, less than]

Table 2   39

Local well name	Sampling date	alpha-Endosulfan, water, filtered, recoverable, in µg/L (34362)	Atrazine, water, filtered, recoverable, in µg/L (39632)	Azinphos-methyl oxygen analog, water, filtered, recoverable, in µg/L (61635)	Azinphos-methyl, water, filtered (0.7 µm glass fiber filter), recoverable, in µg/L (82686)	Benfluralin, water, filtered (0.7 µm glass fiber filter), recoverable, in µg/L (82673)	Bromo-methane, water, unfiltered, recoverable, in µg/L (34413)	Carbaryl, water, filtered (0.7 µm glass fiber filter), recoverable, in µg/L (82680)	Carbofuran, water, filtered (0.7 µm glass fiber filter), recoverable, in µg/L (82674)	Carbon disulfide, water, unfiltered, in µg/L (77041)	Chlorpyrifos oxygen analog, water, filtered, recoverable, in µg/L (61636)
East Basin	2/8/2006	—	e0.005	<0.070	<0.050	<0.010	<0.26	<0.041	—	<0.04	<0.06
Seep A	2/6/2006	—	e0.005	<0.070	<0.050	<0.010	<0.26	<0.041	—	<0.04	<0.06
Seep E	2/8/2006	—	e0.005	<0.070	<0.050	<0.010	<0.26	<0.041	—	<0.04	<0.06
TB–17S	2/7/2006	—	e0.005	<0.070	<0.050	<0.010	<0.26	<0.041	—	<0.04	<0.06
TB–5S	5/1/2006	<0.011	e0.004	<0.042	<0.050	<0.010	<0.33	<0.041	<0.020	<0.04	<0.06
B–3P	5/1/2006	<0.011	e0.007	<0.042	<0.050	<0.010	<0.33	<0.041	<0.020	<0.04	<0.06
TB–12	5/1/2006	<0.011	e0.005	<0.042	<0.050	<0.010	<0.33	<0.041	<0.020	<0.04	<0.06
TB–15	5/2/2006	<0.011	e0.004	<0.042	<0.050	<0.010	<0.33	<0.041	<0.020	<0.04	<0.06
TB–18D	5/1/2006	<0.011	e0.004	<0.042	<0.050	<0.010	<0.33	<0.041	<0.020	0.37	<0.06
TB–1S	2/7/2006	—	<0.007	<0.070	<0.050	<0.010	<0.26	<0.041	—	e0.08	<0.06
TB–18S	5/2/2006	<0.011	<0.007	<0.042	<0.050	<0.010	<0.33	<0.041	<0.020	<0.04	<0.06
TB–1D	2/7/2006	—	e0.005	<0.070	<0.050	<0.010	<0.26	<0.041	—	<0.04	<0.06
TB–5D	2/6/2006	—	e0.005	<0.070	<0.050	<0.010	<0.26	<0.041	—	<0.04	<0.06
TB–2D	2/7/2006	—	<0.007	<0.070	<0.050	<0.010	<0.26	<0.041	—	<0.04	<0.06
TB–17D	2/7/2006	—	<0.007	<0.070	<0.050	<0.010	<0.26	<0.041	—	e0.05	<0.06

Table 2. Concentrations of detected constituents in water samples collected from the Hillview Reservoir, City of Yonkers, Westchester County, New York, in 2006.—Continued

[Numbers in parentheses are National Water Information System (NWIS) parameter codes. DCPA, dimethyl tetrachloroterephthalate; e, estimated; EPTC, S-ethyl dipropylthiocarbamate; BLS, below land surface datum; mg/L, milligrams per liter; µg/L, micrograms per liter; µS/cm, microsiemens per centimeter at 25 degrees Celsius (C); N, nitrogen; P, phosphorus; SiO_2, silicon dioxide; µm, micrometer; <, less than]

Local well name	Sampling date	Chlorpyrifos, water, filtered recoverable, in µg/L (38933)	cis-1,3-Dichloropropene, water, unfiltered, recoverable, in µg/L (34704)	cis-Permethrin, water, filtered (0.7 µm glass fiber filter), recoverable, in µg/L (82687)	cis-Propiconazole, water, filtered, recoverable, in µg/L (79846)	Cyanazine, water, filtered, recoverable, in µg/L (04041)	Cyfluthrin, water, filtered, recoverable, in µg/L (61585)	Cypermethrin, water, filtered, recoverable, in µg/L (61586)	DCPA, water, filtered (0.7 µm glass fiber filter), recoverable, in µg/L (82682)	Desulfinyl-fipronil amide, water, filtered, recoverable, in µg/L (62169)	Desulfinyl-fipronil, water, filtered, recoverable, in µg/L (62170)
East Basin	2/8/2006	<0.0050	<0.05	<0.006	—	—	<0.027	<0.009	<0.0030	<0.029	<0.012
Seep A	2/6/2006	<0.0050	<0.05	<0.006	—	—	<0.027	<0.009	<0.0030	<0.029	<0.012
Seep E	2/8/2006	<0.0050	<0.05	<0.006	—	—	<0.027	<0.009	<0.0030	<0.029	<0.012
TB-17S	2/7/2006	<0.0050	<0.05	<0.006	—	—	<0.027	<0.009	<0.0030	<0.029	<0.012
TB-5S	5/1/2006	<0.0050	<0.05	<0.006	<0.013	<0.018	<0.053	<0.046	<0.0030	<0.029	<0.012
B-3P	5/1/2006	<0.0050	<0.05	<0.006	<0.013	<0.018	<0.053	<0.046	<0.0030	<0.029	<0.012
TB-12	5/1/2006	<0.0050	<0.05	<0.006	<0.013	<0.018	<0.053	<0.046	<0.0030	<0.029	<0.012
TB-15	5/2/2006	<0.0050	<0.05	<0.006	<0.013	<0.018	<0.053	<0.046	<0.0030	<0.029	<0.012
TB-18D	5/1/2006	<0.0050	<0.05	<0.006	<0.013	<0.018	<0.053	<0.046	<0.0030	<0.029	<0.012
TB-1S	2/7/2006	<0.0050	<0.05	<0.006	—	—	<0.027	<0.009	<0.0030	<0.029	e0.004
TB-18S	5/2/2006	<0.0050	<0.05	<0.006	<0.013	<0.018	<0.053	<0.046	<0.0030	<0.029	<0.012
TB-1D	2/7/2006	<0.0050	<0.05	<0.006	—	—	<0.027	<0.009	<0.0030	<0.029	<0.012
TB-5D	2/6/2006	<0.0050	<0.05	<0.006	—	—	<0.027	<0.009	<0.0030	<0.029	<0.012
TB-2D	2/7/2006	<0.0050	<0.05	<0.006	—	—	<0.027	<0.009	<0.0030	<0.029	<0.012
TB-17D	2/7/2006	<0.0050	<0.05	<0.006	—	—	<0.027	<0.009	<0.0030	<0.029	<0.012

Table 2   41

Table 2. Concentrations of detected constituents in water samples collected from the Hillview Reservoir, City of Yonkers, Westchester County, New York, in 2006.—Continued

[Numbers in parentheses are National Water Information System (NWIS) parameter codes. DCPA, dimethyl tetrachloroterephthalate; e, estimated; EPTC, S-ethyl dipropylthiocarbamate; BLS, below land surface datum; mg/L, milligrams per liter; μg/L, micrograms per liter; μS/cm, microsiemens per centimeter at 25 degrees Celsius (C); N, nitrogen; P, phosphorus; SiO$_2$, silicon dioxide; μm, micrometer; <, less than]

Local well name	Sampling date	Diazinon, water, filtered, recoverable, in μg/L (39572)	Dichlorvos, water, filtered, recoverable, in μg/L (38775)	Dicrotophos, water, filtered, recoverable, in μg/L (38454)	Dieldrin, water, filtered, recoverable, in μg/L (39381)	Dimethoate, water, filtered (0.7 μm glass fiber filter), recoverable, in μg/L (82662)	Disulfoton sulfone, water, filtered, recoverable, in μg/L (61640)	Disulfoton, water, filtered (0.7 μm glass fiber filter), recoverable, in μg/L (82677)	Endosulfan sulfate, water, filtered, recoverable, in μg/L (61590)	EPTC, water, filtered (0.7 μm glass fiber filter), recoverable, in μg/L (82668)	Ethion monoxon, water, filtered, recoverable, in μg/L (61644)
East Basin	2/8/2006	<0.0050	<0.01	<0.08	<0.009	<0.0061	—	—	—	—	<0.002
Seep A	2/6/2006	<0.0050	<0.01	<0.08	<0.009	<0.0061	—	—	—	—	<0.002
Seep E	2/8/2006	<0.0050	<0.01	<0.08	<0.009	<0.0061	—	—	—	—	<0.002
TB–17S	2/7/2006	<0.0050	<0.01	<0.08	<0.009	<0.0061	—	—	—	—	<0.002
TB–5S	5/1/2006	<0.0050	<0.01	<0.08	<0.009	<0.0061	<0.014	<0.021	<0.022	<0.0040	<0.021
B–3P	5/1/2006	<0.0050	<0.01	<0.08	<0.009	<0.0061	<0.014	<0.021	<0.022	<0.0040	<0.021
TB–12	5/1/2006	<0.0050	<0.01	<0.08	<0.009	<0.0061	<0.014	<0.021	<0.022	<0.0040	<0.021
TB–15	5/2/2006	<0.0050	<0.01	<0.08	<0.009	<0.0061	<0.014	<0.021	<0.022	<0.0040	<0.021
TB–18D	5/1/2006	<0.0050	<0.01	<0.08	<0.009	<0.0061	<0.014	<0.021	<0.022	<0.0040	<0.021
TB–1S	2/7/2006	<0.0050	<0.01	<0.08	<0.009	<0.0061	—	—	—	—	<0.002
TB–18S	5/2/2006	<0.0050	<0.01	<0.08	<0.009	<0.0061	<0.014	<0.021	<0.022	<0.0040	<0.021
TB–1D	2/7/2006	<0.0050	<0.01	<0.08	<0.009	<0.0061	—	—	—	—	<0.002
TB–5D	2/6/2006	<0.0050	<0.01	<0.08	<0.009	<0.0061	—	—	—	—	<0.002
TB–2D	2/7/2006	<0.0050	<0.01	<0.08	<0.009	<0.0061	—	—	—	—	<0.002
TB–17D	2/7/2006	<0.0050	<0.01	<0.08	<0.009	<0.0061	—	—	—	—	<0.002

Table 2. Concentrations of detected constituents in water samples collected from the Hillview Reservoir, City of Yonkers, Westchester County, New York, in 2006.—Continued

[Numbers in parentheses are National Water Information System (NWIS) parameter codes. DCPA, dimethyl tetrachloroterephthalate; e, estimated; EPTC, S-ethyl dipropylthiocarbamate; BLS, below land surface datum; mg/L, milligrams per liter; µg/L, micrograms per liter; µS/cm, microsiemens per centimeter at 25 degrees Celsius (C); N, nitrogen; P, phosphorus; SiO_2, silicon dioxide; µm, micrometer; <, less than]

Local well name	Sampling date	Ethion, water, filtered recoverable, in µg/L (82346)	Ethoprop, water, filtered (0.7 µm glass fiber filter), recoverable, in µg/L (82672)	Fenamiphos sulfone, water, filtered, recoverable, in µg/L (61645)	Fenamiphos sulfoxide, water, filtered, recoverable, in µg/L (61646)	Fenamiphos, water, filtered, recoverable, in µg/L (61591)	Fipronil sulfide, water, filtered, recoverable, in µg/L (62167)	Fipronil sulfone, water, filtered, recoverable, in µg/L (62168)	Fipronil, water, filtered, recoverable, in µg/L (62166)	Fonofos, water, filtered, recoverable, in µg/L (04095)	Hexazinone, water, filtered, recoverable, in µg/L (04025)
East Basin	2/8/2006	<0.004	—	<0.049	<0.04	<0.029	<0.013	<0.024	<0.016	<0.0030	<0.013
Seep A	2/6/2006	<0.004	—	<0.049	<0.04	<0.029	<0.013	<0.024	<0.016	<0.0030	<0.013
Seep E	2/8/2006	<0.004	—	<0.049	<0.04	<0.029	<0.013	<0.024	<0.016	<0.0030	<0.013
TB–17S	2/7/2006	<0.004	—	<0.049	<0.04	<0.029	<0.013	<0.024	<0.016	<0.0030	<0.013
TB–5S	5/1/2006	<0.016	<0.012	<0.053	<0.04	<0.029	<0.013	<0.024	<0.016	<0.0053	<0.026
B–3P	5/1/2006	<0.016	<0.012	<0.053	<0.04	<0.029	<0.013	<0.024	<0.016	<0.0053	<0.026
TB–12	5/1/2006	<0.016	<0.012	<0.053	<0.04	<0.029	<0.013	<0.024	<0.016	<0.0053	<0.026
TB–15	5/2/2006	<0.016	<0.012	<0.053	<0.04	<0.029	<0.013	<0.024	<0.016	<0.0053	<0.026
TB–18D	5/1/2006	<0.016	<0.012	<0.053	<0.04	<0.029	<0.013	<0.024	<0.016	<0.0053	<0.026
TB–1S	2/7/2006	<0.004	—	<0.049	<0.04	<0.029	e0.006	<0.024	<0.016	<0.0030	<0.013
TB–18S	5/2/2006	<0.016	<0.012	<0.053	<0.04	<0.029	<0.013	<0.024	<0.016	<0.0053	<0.026
TB–1D	2/7/2006	<0.004	—	<0.049	<0.04	<0.029	<0.013	<0.024	<0.016	<0.0030	<0.013
TB–5D	2/6/2006	<0.004	—	<0.049	<0.04	<0.029	<0.013	<0.024	<0.016	<0.0030	<0.013
TB–2D	2/7/2006	<0.004	—	<0.049	<0.04	<0.029	<0.013	<0.024	<0.016	<0.0030	<0.013
TB–17D	2/7/2006	<0.004	—	<0.049	<0.04	<0.029	<0.013	<0.024	<0.016	<0.0030	<0.013

Table 2   43

Table 2. Concentrations of detected constituents in water samples collected from the Hillview Reservoir, City of Yonkers, Westchester County, New York, in 2006.—Continued

[Numbers in parentheses are National Water Information System (NWIS) parameter codes. DCPA, dimethyl tetrachloroterephthalate; e, estimated; EPTC, S-ethyl dipropylthiocarbamate; BLS, below land surface datum; mg/L, milligrams per liter; µg/L, micrograms per liter; µS/cm, microsiemens per centimeter at 25 degrees Celsius (C); N, nitrogen; P, phosphorus; SiO2, silicon dioxide; µm, micrometer; <, less than]

Local well name	Sampling date	Iodomethane, water, unfiltered, recoverable, in µg/L (77424)	Iprodione, water, filtered, recoverable, in µg/L (61593)	Isofenphos, water, filtered, recoverable, in µg/L (61594)	lambda-Cyhalothrin, water, filtered, recoverable, in µg/L (61595)	Malaoxon, water, filtered, recoverable, in µg/L (61652)	Malathion, water, filtered, recoverable, in µg/L (39532)	Metalaxyl, water, filtered, recoverable, in µg/L (61596)	Methidathion, water, filtered, recoverable, in µg/L (61598)	Methyl paraoxon, water, filtered, recoverable, in µg/L (61664)	Methyl parathion, water, filtered (0.7 µm glass fiber filter), recoverable, in µg/L (82667)
East Basin	2/8/2006	<0.50	<0.538	<0.003	—	<0.030	<0.027	<0.005	<0.006	<0.030	<0.015
Seep A	2/6/2006	<0.50	<0.538	<0.003	—	<0.030	<0.027	<0.005	<0.006	<0.030	<0.015
Seep E	2/8/2006	<0.50	<0.538	<0.003	—	<0.030	<0.027	<0.005	<0.006	<0.030	<0.015
TB-17S	2/7/2006	<0.50	<0.538	<0.003	—	<0.030	<0.027	<0.005	<0.006	<0.030	<0.015
TB-5S	5/1/2006	<0.50	<0.026	<0.011	<0.014	<0.039	<0.027	<0.012	<0.007	<0.019	<0.015
B-3P	5/1/2006	<0.50	<0.026	<0.011	<0.014	<0.039	<0.027	<0.007	<0.009	<0.019	<0.015
TB-12	5/1/2006	<0.50	<0.026	<0.011	<0.014	<0.039	<0.027	<0.008	<0.017	<0.019	<0.015
TB-15	5/2/2006	<0.50	<0.026	<0.011	<0.014	<0.039	<0.027	<0.007	<0.009	<0.019	<0.015
TB-18D	5/1/2006	<0.50	<0.026	<0.011	<0.014	<0.039	<0.027	<0.007	<0.009	<0.019	<0.015
TB-1S	2/7/2006	<0.50	<0.538	<0.003	—	<0.030	<0.027	<0.008	<0.006	<0.030	<0.015
TB-18S	5/2/2006	<0.50	<0.026	<0.011	<0.014	<0.039	<0.027	<0.007	<0.009	<0.019	<0.015
TB-1D	2/7/2006	<0.50	<0.538	<0.003	—	<0.030	<0.027	<0.005	<0.006	<0.030	<0.015
TB-5D	2/6/2006	<0.50	<0.538	<0.003	—	<0.030	<0.027	<0.005	<0.006	<0.030	<0.015
TB-2D	2/7/2006	<0.50	<0.538	<0.003	—	<0.030	<0.027	<0.005	<0.006	<0.030	<0.015
TB-17D	2/7/2006	<0.50	<0.538	<0.003	—	<0.030	<0.027	<0.005	<0.006	<0.030	<0.015

Table 2. Concentrations of detected constituents in water samples collected from the Hillview Reservoir, City of Yonkers, Westchester County, New York, in 2006.—Continued

[Numbers in parentheses are National Water Information System (NWIS) parameter codes. DCPA, dimethyl tetrachloroterephthalate; e, estimated; EPTC, S-ethyl dipropylthiocarbamate; BLS, below land surface datum; mg/L, milligrams per liter; µg/L, micrograms per liter; µS/cm, microsiemens per centimeter at 25 degrees Celsius (C); N, nitrogen; P, phosphorus; SiO₂, silicon dioxide; µm, micrometer; <, less than]

Local well name	Sampling date	Metolachlor, water, filtered, recoverable, in µg/L (39415)	Metribuzin, water, filtered, recoverable, in µg/L (82630)	Molinate, water, filtered (0.7 µm glass fiber filter), recoverable, in µg/L (82671)	Myclobutanil, water, filtered, recoverable, in µg/L (61599)	Oxyfluorfen, water, filtered, recoverable, in µg/L (61600)	Pendimethalin, water, filtered (0.7 µm glass fiber filter), recoverable, in µg/L (82683)	Phorate oxygen analog, water, filtered, recoverable, in µg/L (61666)	Phorate, water, filtered (0.7 µm glass fiber filter), recoverable, in µg/L (82664)	Phosmet oxygen analog, water, filtered, recoverable, in µg/L (61668)	Phosmet, water, filtered, recoverable, in µg/L (61601)
East Basin	2/8/2006	e0.005	<0.006	—	<0.008	—	<0.022	<0.105	<0.011	—	—
Seep A	2/6/2006	e0.005	<0.006	—	<0.008	—	<0.022	<0.105	<0.011	—	—
Seep E	2/8/2006	e0.005	<0.006	—	<0.008	—	<0.022	<0.105	<0.011	—	—
TB-17S	2/7/2006	e0.005	<0.006	—	<0.008	—	<0.022	<0.105	<0.011	—	—
TB-5S	5/1/2006	<0.006	<0.028	<0.0030	<0.033	<0.017	<0.022	<0.027	<0.055	<0.0511	<0.008
B-3P	5/1/2006	<0.006	<0.028	<0.0030	<0.033	<0.017	<0.022	<0.027	<0.055	<0.0511	<0.008
TB-12	5/1/2006	<0.006	<0.028	<0.0030	<0.033	<0.017	<0.022	<0.027	<0.055	<0.0511	<0.008
TB-15	5/2/2006	<0.006	<0.028	<0.0030	<0.033	<0.017	<0.022	<0.027	<0.055	<0.0511	<0.008
TB-18D	5/1/2006	<0.006	<0.028	<0.0030	<0.033	<0.017	<0.022	<0.027	<0.055	<0.0511	<0.008
TB-1S	2/7/2006	<0.006	<0.006	—	<0.008	—	<0.022	<0.105	<0.011	—	—
TB-18S	5/2/2006	<0.006	<0.028	<0.0030	<0.033	<0.017	<0.022	<0.027	<0.055	<0.0511	<0.008
TB-1D	2/7/2006	<0.006	<0.006	—	<0.008	—	<0.022	<0.105	<0.011	—	—
TB-5D	2/6/2006	<0.006	<0.006	—	<0.008	—	<0.022	<0.105	<0.011	—	—
TB-2D	2/7/2006	<0.006	<0.006	—	<0.008	—	<0.022	<0.105	<0.011	—	—
TB-17D	2/7/2006	<0.006	<0.006	—	<0.008	—	<0.022	<0.105	<0.011	—	—

Table 2. Concentrations of detected constituents in water samples collected from the Hillview Reservoir, City of Yonkers, Westchester County, New York, in 2006.—Continued

[Numbers in parentheses are National Water Information System (NWIS) parameter codes. DCPA, dimethyl tetrachloroterephthalate; e, estimated; EPTC, S-ethyl dipropylthiocarbamate; BLS, below land surface datum; mg/L, milligrams per liter; µg/L, micrograms per liter; µS/cm, microsiemens per centimeter at 25 degrees Celsius (C); N, nitrogen; P, phosphorus; SiO$_2$, silicon dioxide; µm, micrometer; <, less than]

Table 2   45

Local well name	Sampling date	Prometon, water, filtered, recoverable, in µg/L (04037)	Prometryn, water, filtered, recoverable, in µg/L (04036)	Propanil, water, filtered (0.7 µm glass fiber filter), recoverable, in µg/L (82679)	Propargite, water, filtered (0.7 µm glass fiber filter), recoverable, in µg/L (82685)	Propyzamide, water, filtered (0.7 µm glass fiber filter), recoverable, in µg/L (82676)	Simazine, water, filtered, recoverable, in µg/L (04035)	Tebuthiuron, water, filtered (0.7 µm glass fiber filter), recoverable, in µg/L (82670)	Tefluthrin, water, filtered, recoverable, in µg/L (61606)	Terbufos oxygen analog sulfone, water, filtered, recoverable, in µg/L (61674)	Terbufos, water, filtered (0.7 µm glass fiber filter), recoverable, in µg/L (82675)
East Basin	2/8/2006	<0.010	<0.005	—	—	<0.0040	e0.007	<0.016	—	<0.068	<0.017
Seep A	2/6/2006	<0.010	<0.005	—	—	<0.0040	e0.006	<0.016	—	<0.068	<0.017
Seep E	2/8/2006	<0.010	<0.005	—	—	<0.0040	e0.006	<0.016	—	<0.068	<0.017
TB-17S	2/7/2006	<0.010	<0.005	—	—	<0.0040	<0.005	<0.016	—	<0.068	<0.017
TB-5S	5/1/2006	<0.010	<0.006	<0.011	<0.023	<0.0040	<0.005	<0.016	<0.003	<0.045	<0.017
B-3P	5/1/2006	<0.010	<0.006	<0.011	<0.023	<0.0040	<0.005	<0.016	<0.003	<0.045	<0.017
TB-12	5/1/2006	<0.010	<0.006	<0.011	<0.023	<0.0040	<0.005	<0.016	<0.003	<0.045	<0.017
TB-15	5/2/2006	<0.010	<0.006	<0.011	<0.023	<0.0040	<0.005	<0.016	<0.003	<0.045	<0.017
TB-18D	5/1/2006	<0.010	<0.006	<0.011	<0.023	<0.0040	<0.005	<0.016	<0.003	<0.045	<0.017
TB-1S	2/7/2006	<0.010	<0.005	—	—	<0.0040	<0.005	<0.016	—	<0.068	<0.017
TB-18S	5/2/2006	<0.010	<0.006	<0.011	<0.023	<0.0040	<0.005	<0.016	<0.003	<0.045	<0.017
TB-1D	2/7/2006	<0.010	<0.005	—	—	<0.0040	e0.007	<0.016	—	<0.068	<0.017
TB-5D	2/6/2006	<0.010	<0.005	—	—	<0.0040	e0.006	<0.016	—	<0.068	<0.017
TB-2D	2/7/2006	<0.010	<0.005	—	—	<0.0040	e0.006	<0.016	—	<0.068	<0.017
TB-17D	2/7/2006	<0.010	<0.005	—	—	<0.0040	<0.005	<0.016	—	<0.068	<0.017

Table 2. Concentrations of detected constituents in water samples collected from the Hillview Reservoir, City of Yonkers, Westchester County, New York, in 2006.—Continued

[Numbers in parentheses are National Water Information System (NWIS) parameter codes. DCPA, dimethyl tetrachloroterephthalate; e, estimated; EPTC, S-ethyl dipropylthiocarbamate; BLS, below land surface datum; mg/L, milligrams per liter; µg/L, micrograms per liter; µS/cm, microsiemens per centimeter at 25 degrees Celsius (C), N, nitrogen; P, phosphorus; SiO$_2$, silicon dioxide; µm, micrometer; <, less than]

Local well name	Sampling date	Terbuthylazine, water, filtered, recoverable, in µg/L (04022)	Thiobencarb, water, filtered (0.7 µm glass fiber filter), recoverable, in µg/L (82681)	trans-1,3-Dichloropropene, water, unfiltered, recoverable, in µg/L (34699)	trans-Propiconazole, water, filtered, recoverable, in µg/L (79847)	Tribuphos, water, filtered, recoverable, in µg/L (61610)	Trifluralin, water, filtered (0.7 µm glass fiber filter), recoverable, in µg/L (82661)	1,1,1,2-Tetrachloroethane, water, unfiltered, recoverable, n µg/L (77562)	1,1,1-Trichloroethane, water, unfiltered, recoverable, in µg/L (34506)	1,1,2,2-Tetrachloroethane, water, unfiltered, recoverable, in µg/L (34516)	1,1,2-Trichloro-1,2,2-trifluoroethane, water, unfiltered, recoverable, in µg/L (77652)
East Basin	2/8/2006	<0.010	—	<0.09	—	—	<0.009	<0.030	<0.032	<0.08	<0.038
Seep A	2/6/2006	<0.010	—	<0.09	—	—	<0.009	<0.030	<0.032	<0.08	<0.038
Seep E	2/8/2006	<0.010	—	<0.09	—	—	<0.009	<0.030	<0.032	<0.08	<0.038
TB-17S	2/7/2006	<0.010	—	<0.09	—	—	<0.009	<0.030	<0.032	<0.08	<0.038
TB-5S	5/1/2006	<0.008	<0.010	<0.09	<0.034	<0.035	<0.009	<0.030	<0.032	<0.08	<0.038
B-3P	5/1/2006	<0.008	<0.010	<0.09	<0.034	<0.035	<0.009	<0.030	<0.032	<0.08	<0.038
TB-12	5/1/2006	<0.008	<0.010	<0.09	<0.034	<0.035	<0.009	<0.030	<0.032	<0.08	<0.038
TB-15	5/2/2006	<0.008	<0.010	<0.09	<0.034	<0.035	<0.009	<0.030	<0.032	<0.08	<0.038
TB-18D	5/1/2006	<0.008	<0.010	<0.09	<0.034	<0.035	<0.009	<0.030	<0.032	<0.08	<0.038
TB-1S	2/7/2006	<0.010	—	<0.09	—	—	<0.009	<0.030	<0.032	<0.08	<0.038
TB-18S	5/2/2006	<0.008	<0.010	<0.09	<0.034	<0.035	<0.009	<0.030	<0.032	<0.08	<0.038
TB-1D	2/7/2006	<0.010	—	<0.09	—	—	<0.009	<0.030	<0.032	<0.08	<0.038
TB-5D	2/6/2006	<0.010	—	<0.09	—	—	<0.009	<0.030	<0.032	<0.08	<0.038
TB-2D	2/7/2006	<0.010	—	<0.09	—	—	<0.009	<0.030	<0.032	<0.08	<0.038
TB-17D	2/7/2006	<0.010	—	<0.09	—	—	<0.009	<0.030	<0.032	<0.08	<0.038

Table 2   47

Table 2.   Concentrations of detected constituents in water samples collected from the Hillview Reservoir, City of Yonkers, Westchester County, New York, in 2006.—Continued

[Numbers in parentheses are National Water Information System (NWIS) parameter codes. DCPA, dimethyl tetrachloroterephthalate; e, estimated; EPTC, S-ethyl dipropylthiocarbamate; BLS, below land surface datum; mg/L, milligrams per liter; µg/L, micrograms per liter; µS/cm, microsiemens per centimeter at 25 degrees Celsius (C); N, nitrogen; P, phosphorus; SiO$_2$, silicon dioxide; µm, micrometer; <, less than]

Local well name	Sampling date	1,1,2-Trichloroethane, water, unfiltered, recoverable, in µg/L (34511)	1,1-Dichloroethane, water, unfiltered, recoverable, in µg/L (34496)	1,1-Dichloroethene, water, unfiltered, recoverable, in µg/L (34501)	1,1-Dichloropropene, water, unfiltered, recoverable, in µg/L (77168)	1,2,3,4-Tetramethyl-benzene, water, unfiltered, recoverable, in µg/L (49999)	1,2,3,5-Tetramethyl-benzene, water, unfiltered, recoverable, in µg/L (50000)	1,2,3-Trichlorobenzene, water, unfiltered, recoverable, in µg/L (77613)	1,2,3-Trimethyl-benzene, water, unfiltered, recoverable, in µg/L (77221)	1,2,4-Trichlorobenzene, water, unfiltered, recoverable, in µg/L (34551)	1,2,4-Trimethyl-benzene, water, unfiltered, recoverable, in µg/L (77222)
East Basin	2/8/2006	<0.040	<0.035	<0.024	<0.026	<0.14	<0.140	<0.18	<0.060	<0.12	<0.056
Seep A	2/6/2006	<0.040	<0.035	<0.024	<0.026	<0.14	<0.140	<0.18	<0.060	<0.12	<0.056
Seep E	2/8/2006	<0.040	<0.035	<0.024	<0.026	<0.14	<0.140	<0.18	<0.060	<0.12	<0.056
TB–17S	2/7/2006	<0.040	<0.035	<0.024	<0.026	<0.14	<0.140	<0.18	<0.060	<0.12	<0.056
TB–5S	5/1/2006	<0.040	<0.035	<0.024	<0.026	<0.14	<0.180	<0.18	<0.090	<0.12	<0.056
B–3P	5/1/2006	<0.040	<0.035	<0.024	<0.026	<0.14	<0.180	<0.18	<0.090	<0.12	<0.056
TB–12	5/1/2006	<0.040	<0.035	<0.024	<0.026	<0.14	<0.180	<0.18	<0.090	<0.12	<0.056
TB–15	5/2/2006	<0.040	<0.035	<0.024	<0.026	<0.14	<0.180	<0.18	<0.090	<0.12	<0.056
TB–18D	5/1/2006	<0.040	<0.035	<0.024	<0.026	<0.14	<0.180	<0.18	<0.090	<0.12	<0.056
TB–1S	2/7/2006	<0.040	<0.035	<0.024	<0.026	<0.14	<0.140	<0.18	<0.060	<0.12	<0.056
TB–18S	5/2/2006	<0.040	<0.035	<0.024	<0.026	<0.14	<0.180	<0.18	<0.090	<0.12	<0.056
TB–1D	2/7/2006	<0.040	<0.035	<0.024	<0.026	<0.14	<0.140	<0.18	<0.060	<0.12	<0.056
TB–5D	2/6/2006	<0.040	<0.035	<0.024	<0.026	<0.14	<0.140	<0.18	<0.060	<0.12	<0.056
TB–2D	2/7/2006	<0.040	<0.035	<0.024	<0.026	<0.14	<0.140	<0.18	<0.060	<0.12	<0.056
TB–17D	2/7/2006	<0.040	<0.035	<0.024	<0.026	<0.14	<0.140	<0.18	<0.060	<0.12	<0.056

Table 2. Concentrations of detected constituents in water samples collected from the Hillview Reservoir, City of Yonkers, Westchester County, New York, in 2006.—Continued

[Numbers in parentheses are National Water Information System (NWIS) parameter codes. DCPA, dimethyl tetrachloroterephthalate; e, estimated; EPTC, S-ethyl dipropylthiocarbamate; BLS, below land surface datum; mg/L, milligrams per liter; µg/L, micrograms per liter; µS/cm, microsiemens per centimeter at 25 degrees Celsius (C); N, nitrogen; P, phosphorus; SiO$_2$, silicon dioxide; µm, micrometer; <, less than]

Local well name	Sampling date	1,2-Dichloro-benzene, water, unfiltered, recoverable, in µg/L (34536)	1,3,5-Trimethyl-benzene, water, unfiltered, recoverable, in µg/L (77226)	1,3-Dichloro-benzene, water, unfiltered, recoverable, in µg/L (34566)	2,2-Dichloro-propane, water, unfiltered, recoverable, in µg/L (77170)	2-Chloro-toluene, water, unfiltered, recoverable, in µg/L (77275)	2-Ethyl-toluene, water, unfiltered, recoverable, in µg/L (77220)	4-Chloro-toluene, water, unfiltered, recoverable, in µg/L (77277)	Isopropyl-toluene, water, unfiltered, recoverable, in µg/L (77356)	Acetone, water, unfiltered, recoverable, in µg/L (81552)	Benzene, water, unfiltered, recoverable, in µg/L (34030)
East Basin	2/8/2006	<0.048	<0.044	<0.030	<0.05	<0.040	<0.060	<0.050	<0.08	<6.0	<0.021
Seep A	2/6/2006	<0.048	<0.044	<0.030	<0.05	<0.040	<0.060	<0.050	<0.08	<6.0	<0.021
Seep E	2/8/2006	<0.048	<0.044	<0.030	<0.05	<0.040	<0.060	<0.050	<0.08	<6.0	<0.021
TB-17S	2/7/2006	<0.048	<0.044	<0.030	<0.05	<0.040	<0.060	<0.050	<0.08	<6.0	<0.021
TB-5S	5/1/2006	<0.048	<0.044	<0.030	<0.05	<0.040	<0.060	<0.050	<0.08	<6.0	<0.021
B-3P	5/1/2006	<0.048	<0.044	<0.030	<0.05	<0.040	<0.060	<0.050	<0.08	<6.0	<0.021
TB-12	5/1/2006	<0.048	<0.044	<0.030	<0.05	<0.040	<0.060	<0.050	<0.08	<6.0	<0.021
TB-15	5/2/2006	<0.048	<0.044	<0.030	<0.05	<0.040	<0.060	<0.050	<0.08	<6.0	<0.021
TB-18D	5/1/2006	<0.048	<0.044	<0.030	<0.05	<0.040	<0.060	<0.050	<0.08	<6.0	<0.021
TB-1S	2/7/2006	<0.048	<0.044	<0.030	<0.05	<0.040	<0.060	<0.050	<0.08	e5.2	<0.021
TB-18S	5/2/2006	<0.048	<0.044	<0.030	<0.05	<0.040	<0.060	<0.050	<0.08	<6.0	<0.021
TB-1D	2/7/2006	<0.048	<0.044	<0.030	<0.05	<0.040	<0.060	<0.050	<0.08	<6.0	<0.021
TB-5D	2/6/2006	<0.048	<0.044	<0.030	<0.05	<0.040	<0.060	<0.050	<0.08	<6.0	<0.021
TB-2D	2/7/2006	<0.048	<0.044	<0.030	<0.05	<0.040	<0.060	<0.050	<0.08	<6.0	<0.021
TB-17D	2/7/2006	<0.048	<0.044	<0.030	<0.05	<0.040	<0.060	<0.050	<0.08	<6.0	<0.021

Table 2   49

Table 2. Concentrations of detected constituents in water samples collected from the Hillview Reservoir, City of Yonkers, Westchester County, New York, in 2006.—Continued

[Numbers in parentheses are National Water Information System (NWIS) parameter codes. DCPA, dimethyl tetrachloroterephthalate; e, estimated; EPTC, S-ethyl dipropylthiocarbamate; BLS, below land surface datum; mg/L, milligrams per liter; μg/L, micrograms per liter; μS/cm, microsiemens per centimeter at 25 degrees Celsius (C); N, nitrogen; P, phosphorus; SiO_2, silicon dioxide; μm, micrometer; <, less than]

Local well name	Sampling date	Bromobenzene, water, unfiltered, recoverable, in μg/L (81555)	Bromochloromethane, water, unfiltered, recoverable, in μg/L (77297)	Bromodichloromethane, water, unfiltered, recoverable, in μg/L (32101)	Bromoethene, water, unfiltered, recoverable, in μg/L (50002)	Chlorobenzene, water, unfiltered, recoverable, in μg/L (34301)	Chloroethane, water, unfiltered, recoverable, in μg/L (34311)	Chloromethane, water, unfiltered, recoverable, in μg/L (34418)	cis-1,2-Dichloroethene, water, unfiltered, recoverable, in μg/L (77093)	Dibromochloromethane, water, unfiltered, recoverable, in μg/L (32105)	Dibromomethane, water, unfiltered, recoverable, in μg/L (30217)
East Basin	2/8/2006	<0.028	<0.12	3.65	<0.10	<0.028	<0.12	<0.170	<0.024	0.27	<0.050
Seep A	2/6/2006	<0.028	<0.12	0.666	<0.10	<0.028	<0.12	<0.170	<0.024	<0.10	<0.050
Seep E	2/8/2006	<0.028	<0.12	0.543	<0.10	<0.028	<0.12	<0.170	<0.024	<0.10	<0.050
TB–17S	2/7/2006	<0.028	<0.12	0.932	<0.10	<0.028	<0.12	<0.170	<0.024	<0.10	<0.050
TB–5S	5/1/2006	<0.028	<0.12	e0.035	<0.10	<0.028	<0.12	<0.170	<0.024	<0.10	<0.050
B–3P	5/1/2006	<0.028	<0.12	<0.028	<0.10	<0.028	<0.12	<0.170	<0.024	<0.10	<0.050
TB–12	5/1/2006	<0.028	<0.12	e0.034	<0.10	<0.028	<0.12	<0.170	<0.024	<0.10	<0.050
TB–15	5/2/2006	<0.028	<0.12	<0.028	<0.10	<0.028	<0.12	<0.170	<0.024	<0.10	<0.050
TB–18D	5/1/2006	<0.028	<0.12	<0.028	<0.10	<0.028	<0.12	<0.170	<0.024	<0.10	<0.050
TB–1S	2/7/2006	<0.028	<0.12	<0.028	<0.10	<0.028	<0.12	<0.170	<0.024	<0.10	<0.050
TB–18S	5/2/2006	<0.028	<0.12	<0.028	<0.10	<0.028	<0.12	<0.170	<0.024	<0.10	<0.050
TB–1D	2/7/2006	<0.028	<0.12	<0.028	<0.10	<0.028	<0.12	<0.170	<0.024	<0.10	<0.050
TB–5D	2/6/2006	<0.028	<0.12	<0.028	<0.10	<0.028	<0.12	<0.170	<0.024	<0.10	<0.050
TB–2D	2/7/2006	<0.028	<0.12	<0.028	<0.10	<0.028	<0.12	<0.170	<0.024	<0.10	<0.050
TB–17D	2/7/2006	<0.028	<0.12	<0.028	<0.10	<0.028	<0.12	<0.170	<0.024	<0.10	<0.050

Table 2. Concentrations of detected constituents in water samples collected from the Hillview Reservoir, City of Yonkers, Westchester County, New York, in 2006.—Continued

[Numbers in parentheses are National Water Information System (NWIS) parameter codes. DCPA, dimethyl tetrachloroterephthalate; e, estimated; EPTC, S-ethyl dipropylthiocarbamate; BLS, below land surface datum; mg/L, milligrams per liter; µg/L, micrograms per liter; µS/cm, microsiemens per centimeter at 25 degrees Celsius (C); N, nitrogen; P, phosphorus; SiO$_2$, silicon dioxide; µm, micrometer; <, less than]

Local well name	Sampling date	Dichlorodifluoromethane, water, unfiltered, recoverable, in µg/L (34668)	Dichloromethane, water, unfiltered, recoverable, in µg/L (34423)	Diethyl ether, water, unfiltered, recoverable, in µg/L (81576)	Diisopropyl ether, water, unfiltered, recoverable, in µg/L (81577)	Ethyl methacrylate, water, unfiltered, recoverable, in µg/L (73570)	Ethyl methyl ketone, water, unfiltered, recoverable, in µg/L (81595)	Ethylbenzene, water, unfiltered, recoverable, in µg/L (34371)	Hexachlorobutadiene, water, unfiltered, recoverable, in µg/L (39702)	Hexachloroethane, water, unfiltered, recoverable, in µg/L (34396)	Isobutyl methyl ketone, water, unfiltered, recoverable, in µg/L (78133)
East Basin	2/8/2006	<0.18	<0.06	<0.1	<0.10	<0.18	<2.0	<0.030	<0.14	<0.14	<0.37
Seep A	2/6/2006	<0.18	e0.04	<0.1	<0.10	<0.18	<2.0	<0.030	<0.14	<0.14	<0.37
Seep E	2/8/2006	<0.18	<0.06	<0.1	<0.10	<0.18	<2.0	<0.030	<0.14	<0.14	<0.37
TB–17S	2/7/2006	<0.18	<0.06	<0.1	<0.10	<0.18	<2.0	<0.030	<0.14	<0.14	<0.37
TB–5S	5/1/2006	<0.18	<0.06	<0.1	<0.10	<0.18	<2.0	<0.030	<0.14	<0.14	<0.37
B–3P	5/1/2006	<0.18	<0.06	<0.1	<0.10	<0.18	<2.0	<0.030	<0.14	<0.14	<0.37
TB–12	5/1/2006	<0.18	<0.06	<0.1	<0.10	<0.18	<2.0	<0.030	<0.14	<0.14	<0.37
TB–15	5/2/2006	<0.18	<0.06	<0.1	<0.10	<0.18	<2.0	<0.030	<0.14	<0.14	<0.37
TB–18D	5/1/2006	<0.18	<0.06	<0.1	<0.10	<0.18	<2.0	<0.030	<0.14	<0.14	<0.37
TB–1S	2/7/2006	<0.18	<0.06	<0.1	<0.10	<0.18	<2.0	<0.030	<0.14	<0.14	<0.37
TB–18S	5/2/2006	<0.18	<0.06	<0.1	<0.10	<0.18	<2.0	<0.030	<0.14	<0.14	<0.37
TB–1D	2/7/2006	<0.18	<0.06	<0.1	<0.10	<0.18	<2.0	<0.030	<0.14	<0.14	<0.37
TB–5D	2/6/2006	<0.18	<0.06	<0.1	<0.10	<0.18	<2.0	<0.030	<0.14	<0.14	<0.37
TB–2D	2/7/2006	<0.18	<0.06	<0.1	<0.10	<0.18	<2.0	<0.030	<0.14	<0.14	<0.37
TB–17D	2/7/2006	<0.18	<0.06	<0.1	<0.10	<0.18	<2.0	<0.030	<0.14	<0.14	<0.37

Table 2   51

Table 2. Concentrations of detected constituents in water samples collected from the Hillview Reservoir, City of Yonkers, Westchester County, New York, in 2006.—Continued

[Numbers in parentheses are National Water Information System (NWIS) parameter codes. DCPA, dimethyl tetrachloroterephthalate; e, estimated; EPTC, S-ethyl dipropylthiocarbamate; BLS, below land surface datum; mg/L, milligrams per liter; µg/L, micrograms per liter; µS/cm, microsiemens per centimeter at 25 degrees Celsius (C); N, nitrogen; P, phosphorus; SiO$_2$, silicon dioxide; µm, micrometer; <, less than]

Local well name	Sampling date	Isopropyl-benzene, water, unfiltered, recoverable, in µg/L (77223)	Methyl acrylate, water, unfiltered, recoverable, in µg/L (49991)	Methyl acrylonitrile, water, unfiltered, recoverable, in µg/L (81593)	Methyl meth-acrylate, water, unfiltered, recoverable, in µg/L (81597)	Methyl tert-butyl ether, water, unfiltered, recoverable, in µg/L (78032)	Methyl tert-pentyl ether, water, unfiltered, recoverable, in µg/L (50005)	Methylene blue active substances, water, unfiltered, recoverable, milligrams per liter (38260)	m-Xylene plus p-xylene, water, unfiltered, recoverable, in µg/L (85795)	Naphthalene, water, unfiltered, recoverable, in µg/L (34696)	n-Butyl methyl ketone, water, unfiltered, recoverable, in µg/L (77103)
East Basin	2/8/2006	<0.038	<1.0	<0.40	<0.20	<0.10	<0.04	<0.100	<0.06	<0.52	<0.4
Seep A	2/6/2006	<0.038	<1.0	<0.40	<0.20	<0.10	<0.04	<0.100	<0.06	<0.52	<0.4
Seep E	2/8/2006	<0.038	<1.0	<0.40	<0.20	<0.10	<0.04	<0.100	<0.06	<0.52	<0.4
TB-17S	2/7/2006	<0.038	<1.0	<0.40	<0.20	<0.10	<0.04	<0.100	e0.03	<0.52	<0.4
TB-5S	5/1/2006	<0.038	<1.0	<0.40	<0.20	<0.10	<0.04	0.121	<0.06	<0.52	<0.4
B-3P	5/1/2006	<0.038	<1.0	<0.40	<0.20	<0.10	<0.04	0.123	<0.06	<0.52	<0.4
TB-12	5/1/2006	<0.038	<1.0	<0.40	<0.20	<0.10	<0.04	0.194	<0.06	<0.52	<0.4
TB-15	5/2/2006	<0.038	<1.0	<0.40	<0.20	<0.10	<0.04	<0.100	<0.06	<0.52	<0.4
TB-18D	5/1/2006	<0.038	<1.0	<0.40	<0.20	<0.10	<0.04	<0.100	<0.06	<0.52	<0.4
TB-1S	2/7/2006	<0.038	<1.0	<0.40	<0.20	<0.10	<0.04	e0.085	<0.06	<0.52	<0.4
TB-18S	5/2/2006	<0.038	<1.0	<0.40	<0.20	<0.10	<0.04	<0.100	<0.06	<0.52	<0.4
TB-1D	2/7/2006	<0.038	<1.0	<0.40	<0.20	<0.10	<0.04	—	<0.06	<0.52	<0.4
TB-5D	2/6/2006	<0.038	<1.0	<0.40	<0.20	<0.10	<0.04	<0.100	<0.06	<0.52	<0.4
TB-2D	2/7/2006	<0.038	<1.0	<0.40	<0.20	<0.10	<0.04	<0.100	<0.06	<0.52	<0.4
TB-17D	2/7/2006	<0.038	<1.0	<0.40	<0.20	<0.10	<0.04	e0.072	<0.06	<0.52	<0.4

Table 2.   Concentrations of detected constituents in water samples collected from the Hillview Reservoir, City of Yonkers, Westchester County, New York, in 2006.—Continued

[Numbers in parentheses are National Water Information System (NWIS) parameter codes. DCPA, dimethyl tetrachloroterephthalate; e, estimated; EPTC, S-ethyl dipropylthiocarbamate; BLS, below land surface datum; mg/L, milligrams per liter; µg/L, micrograms per liter; µS/cm, microsiemens per centimeter at 25 degrees Celsius (O); N, nitrogen; P, phosphorus; SiO$_2$, silicon dioxide; µm, micrometer; <, less than]

Local well name	Sampling date	n-Butyl-benzene, water, unfiltered, recoverable, in µg/L (77342)	n-Propyl-benzene, water, unfiltered, recoverable, in µg/L (77224)	o-Xylene, water, unfiltered, recoverable, in µg/L (77135)	sec-Butyl-benzene, water, unfiltered, recoverable, in µg/L (77350)	Styrene, water, unfiltered, recoverable, in µg/L (77128)	tert-Butyl ethyl ether, water, unfiltered, recoverable, in µg/L (50004)	tert-Butyl-benzene, water, unfiltered, recoverable, in µg/L (77353)	Tetrachloro-ethene, water, unfiltered, recoverable, in µg/L (34475)	Tetrachloro-methane, water, unfiltered, recoverable, in µg/L (32102)	Tetra-hydrofuran, water, unfiltered, recoverable, in µg/L (81607)
East Basin	2/8/2006	<0.12	<0.042	<0.038	<0.060	<0.042	<0.030	<0.060	<0.030	<0.06	<1.0
Seep A	2/6/2006	<0.12	<0.042	<0.038	<0.060	<0.042	<0.030	<0.060	<0.030	<0.06	<1.0
Seep E	2/8/2006	<0.12	<0.042	<0.038	<0.060	<0.042	<0.030	<0.060	<0.030	<0.06	<1.0
TB-17S	2/7/2006	<0.12	<0.042	<0.038	<0.060	<0.042	<0.030	<0.060	<0.030	<0.06	<1.0
TB-5S	5/1/2006	<0.14	<0.042	<0.038	<0.060	<0.042	<0.030	<0.080	<0.030	<0.06	<1.2
B-3P	5/1/2006	<0.14	<0.042	<0.038	<0.060	<0.042	<0.030	<0.080	<0.030	<0.06	<1.2
TB-12	5/1/2006	<0.14	<0.042	<0.038	<0.060	<0.042	<0.030	<0.080	<0.030	<0.06	<1.2
TB-15	5/2/2006	<0.14	<0.042	<0.038	<0.060	<0.042	<0.030	<0.080	<0.030	<0.06	<1.2
TB-18D	5/1/2006	<0.14	<0.042	<0.038	<0.060	<0.042	<0.030	<0.080	<0.030	<0.06	<1.2
TB-1S	2/7/2006	<0.12	<0.042	<0.038	<0.060	<0.042	<0.030	<0.060	<0.030	<0.06	<1.0
TB-18S	5/2/2006	<0.14	<0.042	<0.038	<0.060	<0.042	<0.030	<0.080	<0.030	<0.06	<1.2
TB-1D	2/7/2006	<0.12	<0.042	<0.038	<0.060	<0.042	<0.030	<0.060	<0.030	<0.06	<1.0
TB-5D	2/6/2006	<0.12	<0.042	<0.038	<0.060	<0.042	<0.030	<0.060	<0.030	<0.06	<1.0
TB-2D	2/7/2006	<0.12	<0.042	<0.038	<0.060	<0.042	<0.030	<0.060	<0.030	<0.06	<1.0
TB-17D	2/7/2006	<0.12	<0.042	<0.038	<0.060	<0.042	<0.030	<0.060	<0.030	<0.06	<1.0

Table 2   53

Table 2. Concentrations of detected constituents in water samples collected from the Hillview Reservoir, City of Yonkers, Westchester County, New York, in 2006.—Continued

[Numbers in parentheses are National Water Information System (NWIS) parameter codes. DCPA, dimethyl tetrachloroterephthalate; e, estimated; EPTC, S-ethyl dipropylthiocarbamate; BLS, below land surface datum; mg/L, milligrams per liter; μg/L, micrograms per liter; μS/cm, microsiemens per centimeter at 25 degrees Celsius (C); N, nitrogen; P, phosphorus; SiO$_2$, silicon dioxide; μm, micrometer; <, less than]

Local well name	Sampling date	Toluene, water, unfiltered, recoverable, in µg/L (34010)	trans-1,2-Dichloro-ethene, water, unfiltered, recoverable, in µg/L (34546)	trans-1,4-Dichloro-2-butene, water, unfiltered, recoverable, in µg/L (73547)	Tribromo-methane, water, unfiltered, recoverable, in µg/L (32104)	Trichloro-ethene, water, unfiltered, recoverable, in µg/L (39180)	Trichloro-fluoro-methane, water, unfiltered, recoverable, in µg/L (34488)	Trichloro-methane, water, unfiltered, recoverable, in µg/L (32106)	Vinyl chloride, water, unfiltered, recoverable, in µg/L (39175)
East Basin	2/8/2006	e0.018	<0.032	<0.7	<0.10	<0.038	<0.08	14.5	<0.08
Seep A	2/6/2006	<0.020	<0.032	<0.7	<0.10	<0.038	<0.08	7.57	<0.08
Seep E	2/8/2006	<0.020	<0.032	<0.7	<0.10	<0.038	<0.08	7.49	<0.08
TB–17S	2/7/2006	0.107	<0.032	<0.7	<0.10	<0.038	<0.08	12	<0.08
TB–5S	5/1/2006	e0.060	<0.032	<0.7	<0.10	<0.038	<0.08	1.75	<0.08
B–3P	5/1/2006	e0.023	<0.032	<0.7	<0.10	<0.038	<0.08	<0.024	<0.08
TB–12	5/1/2006	e0.012	<0.032	<0.7	<0.10	<0.038	<0.08	1.54	<0.08
TB–15	5/2/2006	<0.020	<0.032	<0.7	<0.10	<0.038	<0.08	14.3	<0.08
TB–18D	5/1/2006	<0.020	<0.032	<0.7	<0.10	<0.038	<0.08	20.9	<0.08
TB–1S	2/7/2006	<0.020	<0.032	<0.7	<0.10	<0.038	<0.08	<0.024	<0.08
TB–18S	5/2/2006	<0.020	<0.032	<0.7	<0.10	<0.038	<0.08	e0.042	<0.08
TB–1D	2/7/2006	<0.020	<0.032	<0.7	<0.10	<0.038	<0.08	22.9	<0.08
TB–5D	2/6/2006	<0.020	<0.032	<0.7	<0.10	<0.038	<0.08	22.4	<0.08
TB–2D	2/7/2006	4.78	<0.032	<0.7	<0.10	<0.038	<0.08	24.1	<0.08
TB–17D	2/7/2006	e0.022	<0.032	<0.7	<0.10	<0.038	<0.08	0.625	<0.08

Appendix 1.—Water Elevation and Precipitation at Selected Wells at the Hillview Reservoir, City of Yonkers, Westchester County, New York

Hydrographs in the following figures show elevation of water levels in relation to daily precipitation at the Hillview Reservoir, City of Yonkers, Westchester County, New York, April 1, 2005–March 1, 2008.

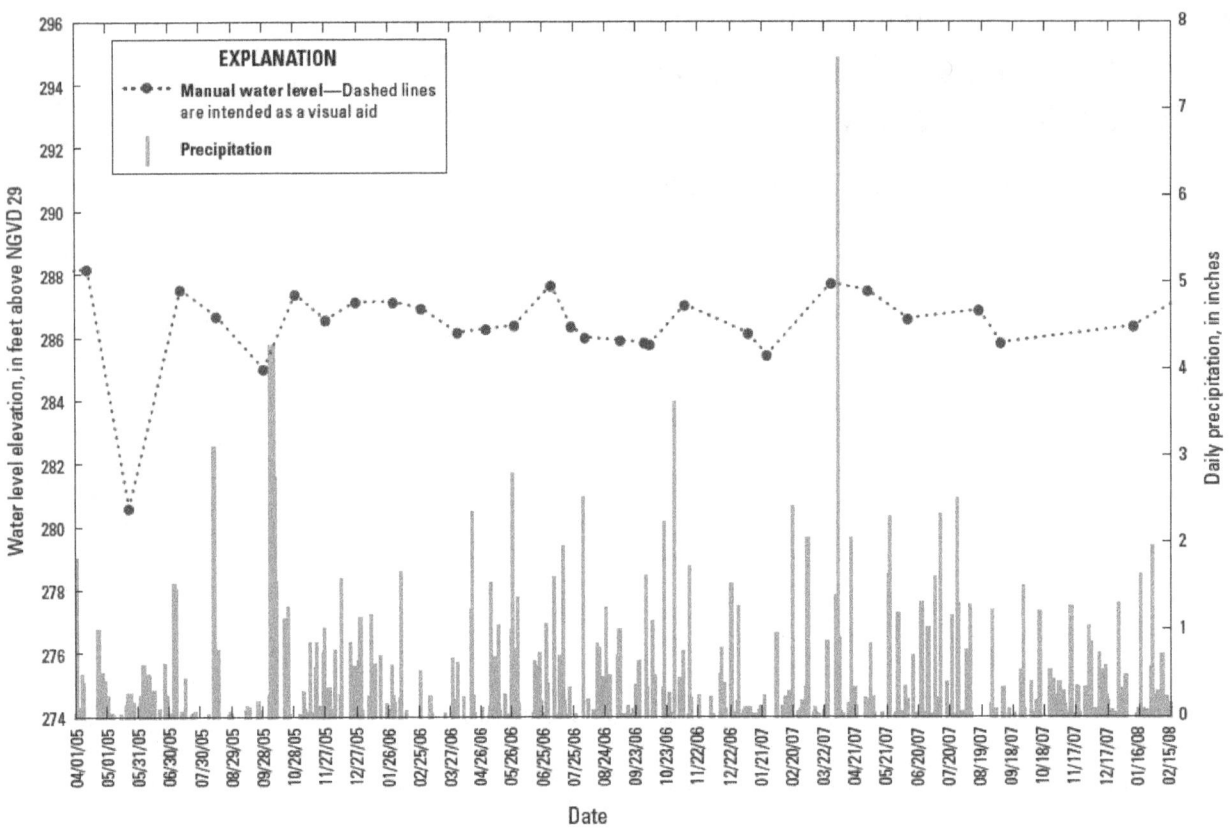

Figure 1–1. Elevation of water levels at well B–3P in relation to daily precipitation at the Hillview Reservoir, City of Yonkers, Westchester County, New York, April 1, 2005–March 1, 2008, as recorded at Central Park (National Oceanic and Atmospheric Administration, 2008). The location of the well is shown in figure 1.

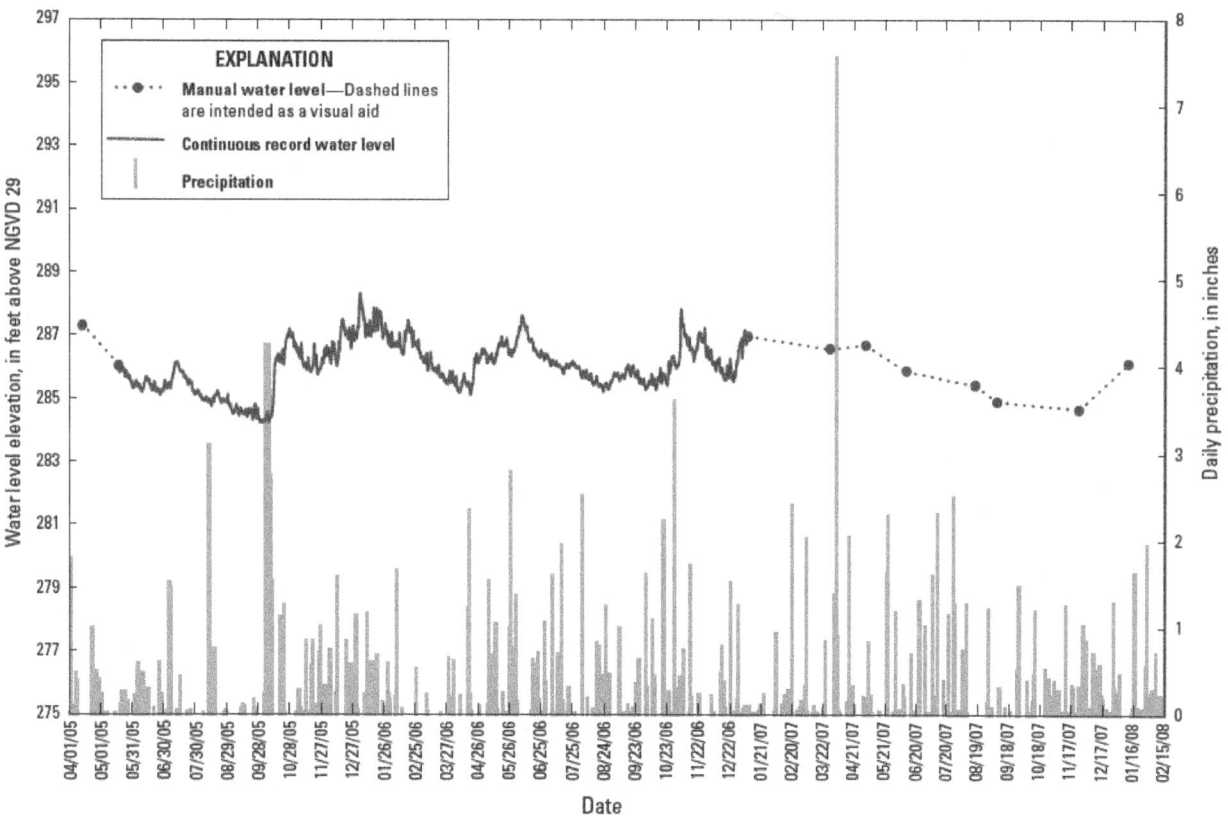

Figure 1–2.   Elevation of water levels at well CMB–2W in relation to daily precipitation at the Hillview Reservoir, City of Yonkers, Westchester County, New York, April 1, 2005–March 1, 2008, as recorded at Central Park (National Oceanic and Atmospheric Administration, 2008). The location of the well is shown in figure 1.

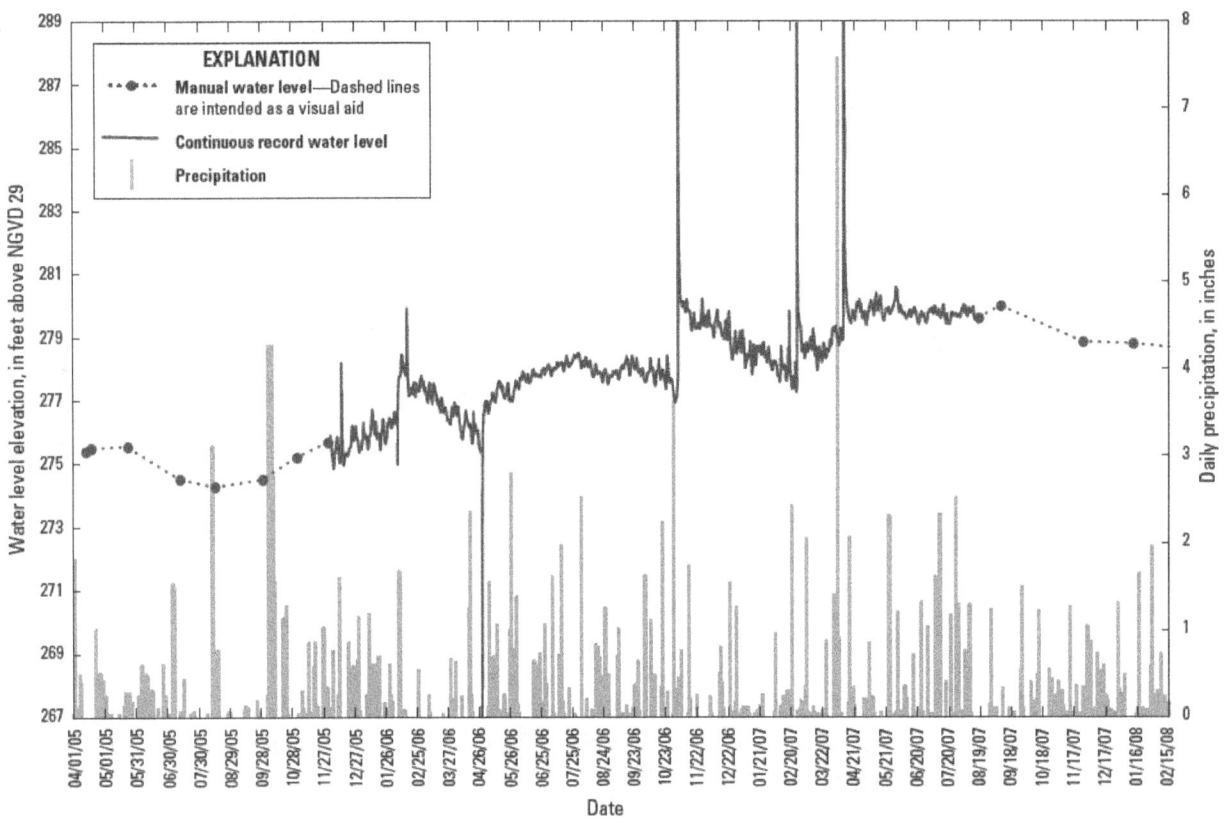

Figure 1–3. Elevation of water levels at well TB–2S in relation to daily precipitation at the Hillview Reservoir, City of Yonkers, Westchester County, New York, April1, 2005–March 1, 2008, as recorded at Central Park (National Oceanic and Atmospheric Administration, 2008). The location of the well is shown in figure 1.

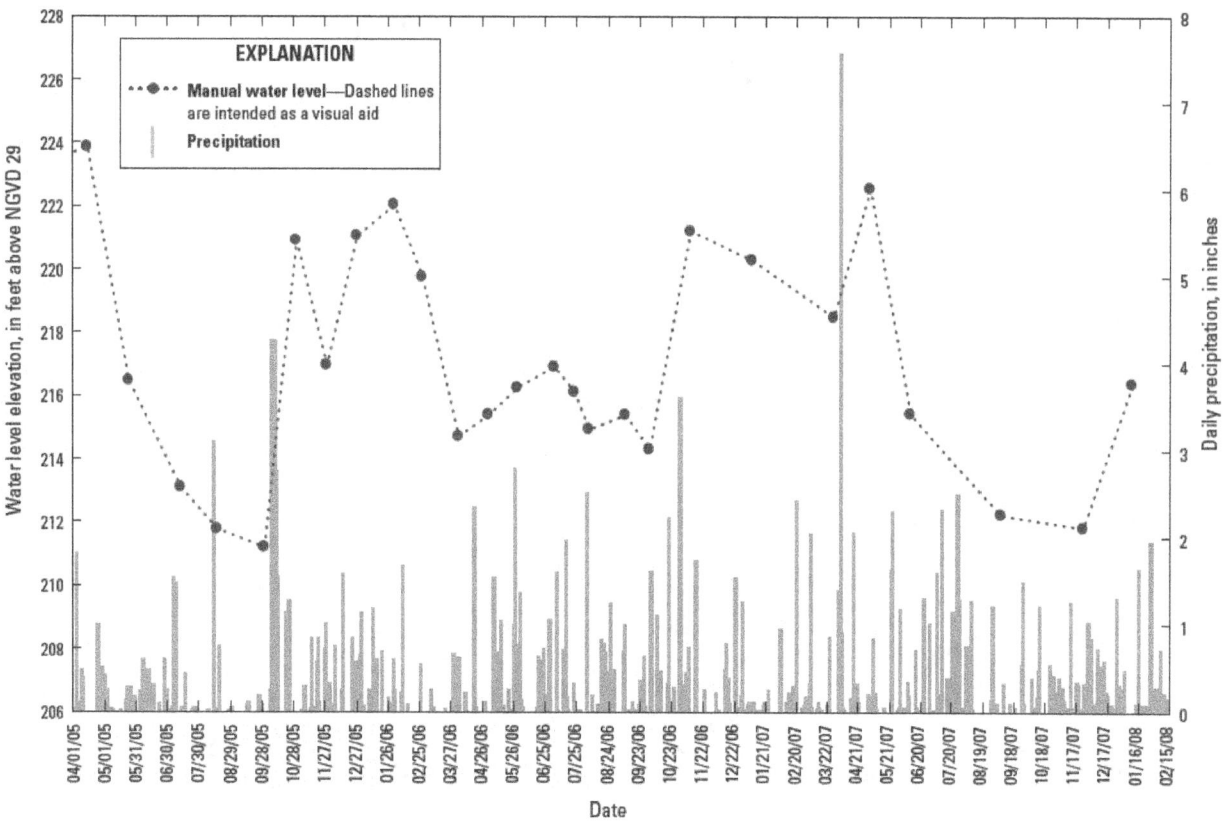

Figure 1–4. Elevation of water levels at well TB–10 in relation to daily precipitation at the Hillview Reservoir, City of Yonkers, Westchester County, New York, April 1, 2005–March 1, 2008, as recorded at Central Park (National Oceanic and Atmospheric Administration, 2008). The location of the well is shown in figure 1.

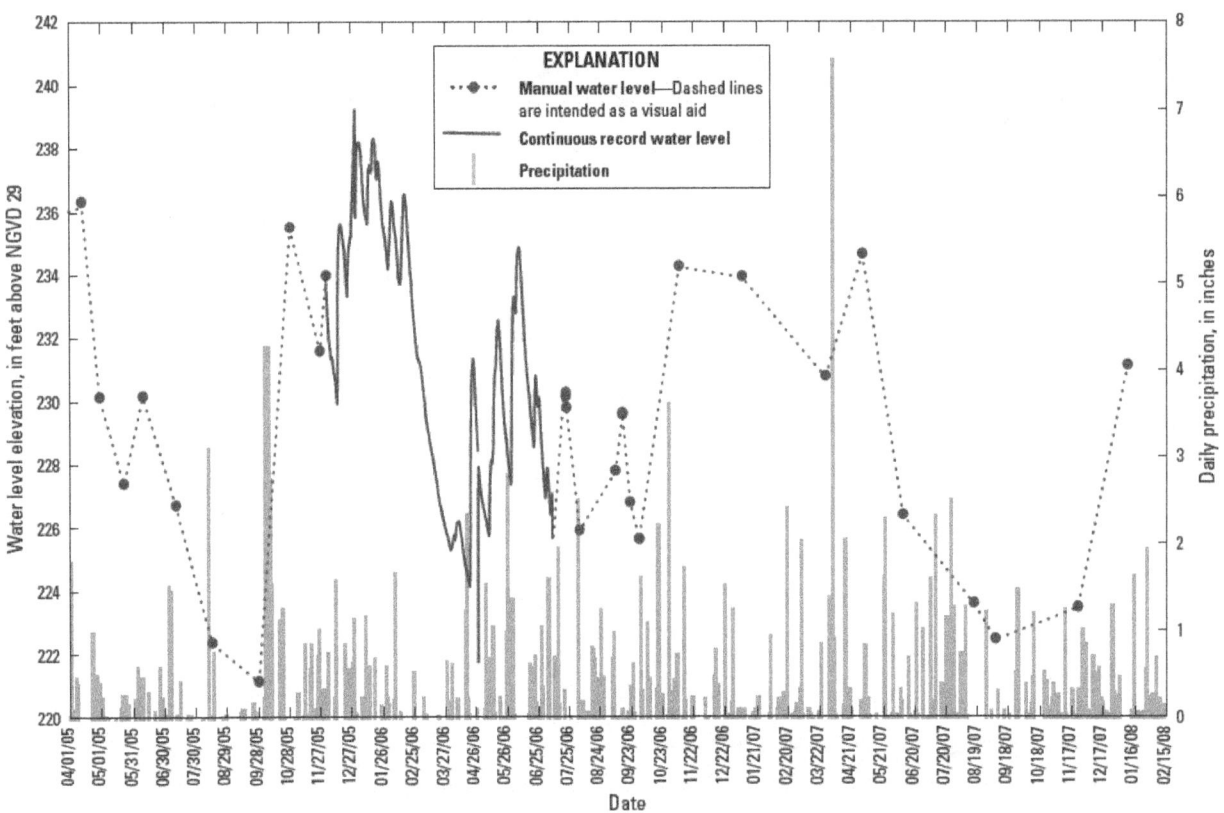

Figure 1–5.   Elevation of water levels at well TB–12 in relation to daily precipitation at the Hillview Reservoir, City of Yonkers, Westchester County, New York, April 1, 2005–March 1, 2008, as recorded at Central Park (National Oceanic and Atmospheric Administration, 2008). The location of the well is shown in figure 1.

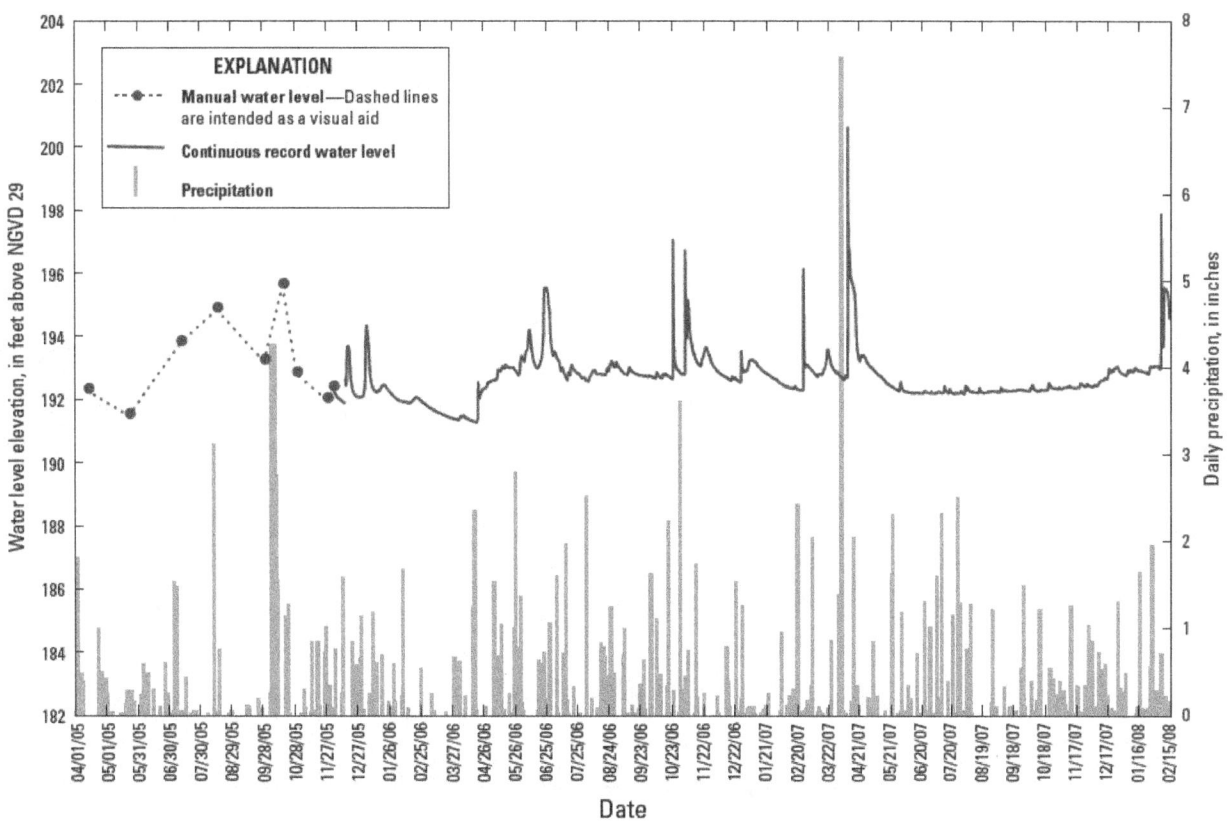

Figure 1–6. Elevation of water levels at well TB–13 in relation to daily precipitation at the Hillview Reservoir, City of Yonkers, Westchester County, New York, April 1, 2005–March 1, 2008, as recorded at Central Park (National Oceanic and Atmospheric Administration, 2008). The location of the well is shown in figure 1.

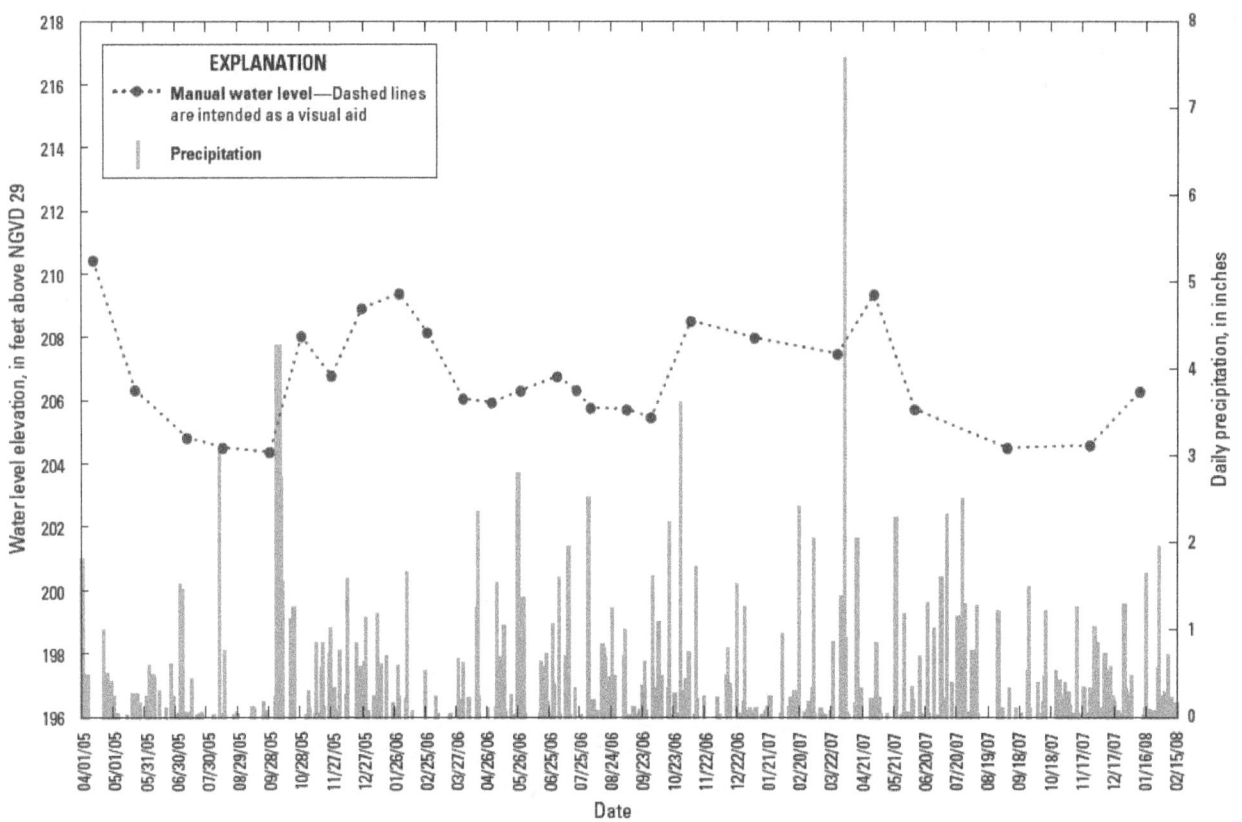

Figure 1–7.   Elevation of water levels at well TB–14D in relation to daily precipitation at the Hillview Reservoir, City of Yonkers, Westchester County, New York, April 1, 2005–March 1, 2008, as recorded at Central Park (National Oceanic and Atmospheric Administration, 2008). The location of the well is shown in figure 1.

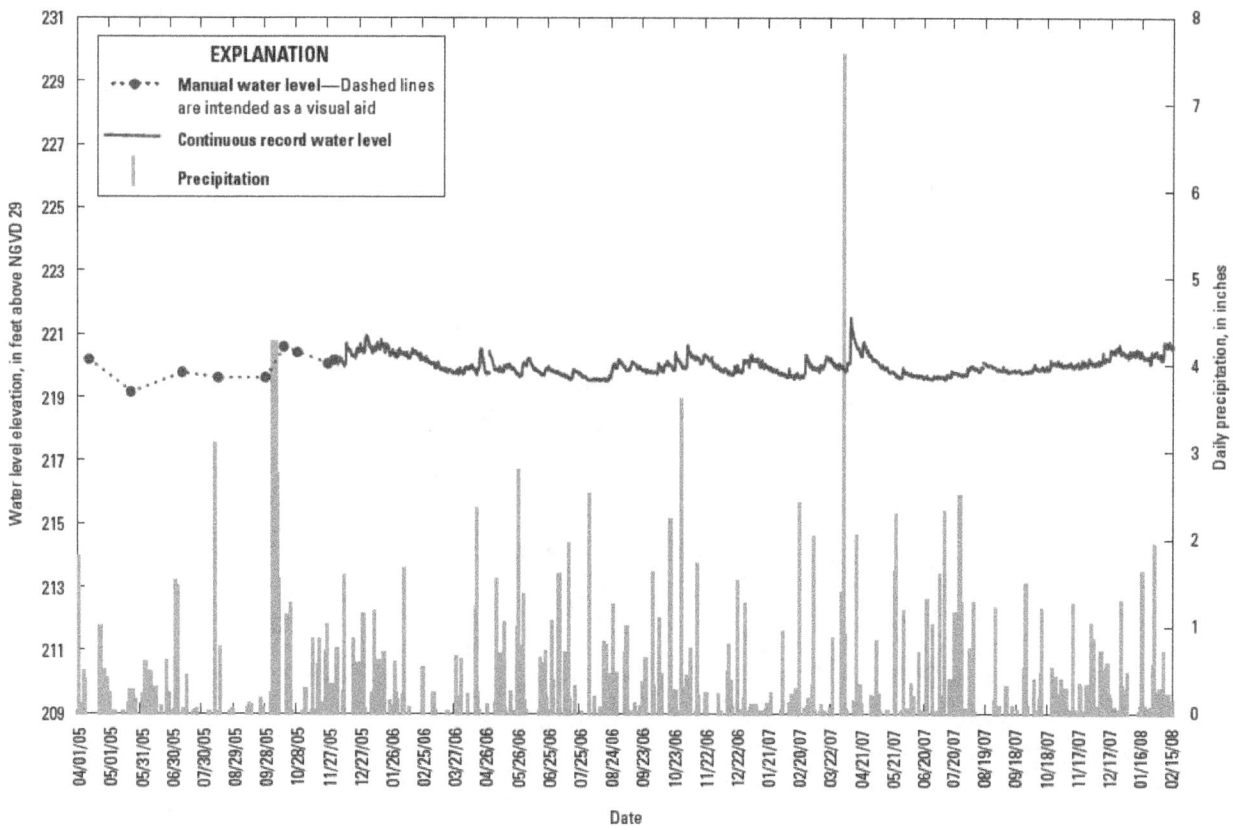

Figure 1–8. Elevation of water levels at well TB–15 in relation to daily precipitation at the Hillview Reservoir, City of Yonkers, Westchester County, New York, April 1, 2005–March 1, 2008, as recorded at Central Park (National Oceanic and Atmospheric Administration, 2008). The location of the well is shown in figure 1.

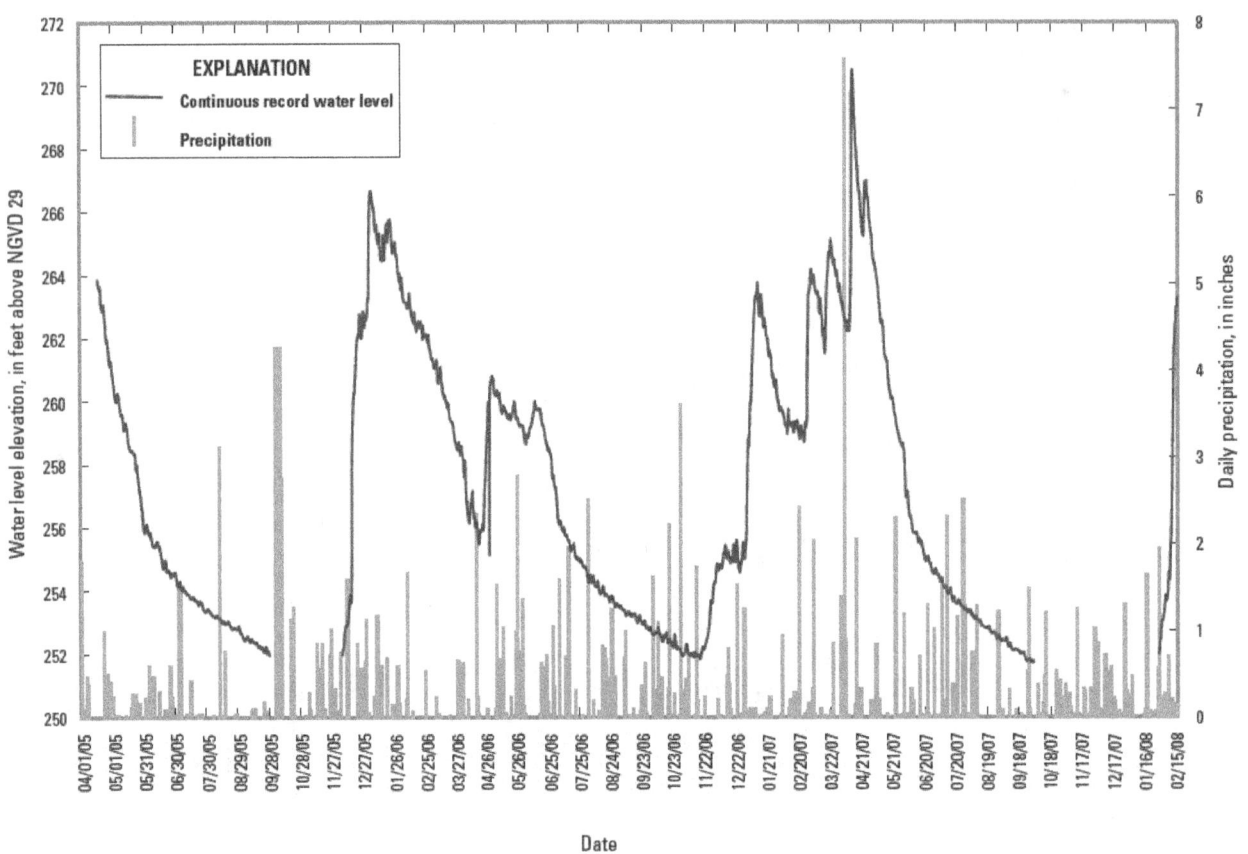

Figure 1–9.   Elevation of water levels at well TB–18S in relation to daily precipitation at the Hillview Reservoir, City of Yonkers, Westchester County, New York, April 1, 2005–March 1, 2008, as recorded at Central Park (National Oceanic and Atmospheric Administration, 2008). Blank where data are missing. The location of the well is shown in figure 1.